POST-DISASTER AND POST-CONFLICT TOURISM

(2nd Edition)

POST-DISASTER AND POST-CONFLICT TOURISM

(2nd Edition)

Edited by

Maximiliano E. Korstanje, PhD

Hugues Séraphin, PhD

Vanessa G. B. Gowreesunkar, PhD

AAP APPLE ACADEMIC PRESS

First edition published 2024

Apple Academic Press Inc.
1265 Goldenrod Circle, NE,
Palm Bay, FL 32905 USA

760 Laurentian Drive, Unit 19,
Burlington, ON L7N 0A4, CANADA

CRC Press
2385 NW Executive Center Drive,
Suite 320, Boca Raton FL 33431

4 Park Square, Milton Park,
Abingdon, Oxon, OX14 4RN UK

Apple Academic Press exclusively co-publishes with CRC Press, an imprint of Taylor & Francis Group, LLC

Library and Archives Canada Cataloguing in Publication

Title: Post-disaster and post-conflict tourism / Edited by Maximiliano E. Korstanje, PhD, Hugues Séraphin, PhD, Vanessa G. B. Gowreesunkar, PhD.
Names: Korstanje, Maximiliano E., editor. | Séraphin, Hugues, editor. | Gowreesunkar, Vanessa G. B., editor.
Description: Second edition. | Includes bibliographical references and index.
Identifiers: Canadiana (print) 20230441661 | Canadiana (ebook) 20230441777 | ISBN 9781774913505 (hardcover) | ISBN 9781774913512 (paperback) | ISBN 9781003387930 (ebook)
Subjects: LCSH: Tourism—Management. | LCSH: Natural disasters—Economic aspects. | LCSH: Tourism—Environmental aspects. | LCSH: Dark tourism.
Classification: LCC G155.A1 P67 2023 | DDC 910.68—dc23

Library of Congress Cataloging-in-Publication Data

Names: Korstanje, Maximiliano E., editor. | Séraphin, Hugues, editor. | Gowreesunkar, Vanessa G. B., editor.

Title: Post-disaster and post-conflict tourism / Edited by Maximiliano E. Korstanje, PhD, Hugues Séraphin, PhD, Vanessa G. B. Gowreesunkar, PhD.

Description: Second edition. | Palm Bay : Apple Academic Press, 2023. | Includes bibliographical references and index. | Summary: "This new volume, Post-Disaster and Post-Conflict Tourism, now going into its 2nd edition, takes an in-depth look at how global geopolitical tensions and global threats affect the tourism industry and offers tools and strategies for meeting these challenges. The book is updated with chapters that include new research, studies, and experiences, many of which consider the fall-out from the COVID-19 pandemic on tourism. It also includes five brand new chapters, for over 50 new pages of text. With chapters by well-versed scholars who have worked as experts in post-disaster and post-conflict tourism, the book presents a host of case scenarios along with innovative strategies that can be implemented by postcolonial, post-conflict, and post-disaster destinations to encourage travel and tourism in these areas. Topics include using tourism as a vehicle for economic recovery, educating tourists at the pre-visit stage, developing and employing postcolonial branding and self-branding, using sports tourism and food events as a marketing strategy, the ethics revolving around post-disaster consumption, and much more. The new chapters discuss tourism in the age of the coronavirus pandemic and its dramatic disruptive effect on the tourism industry. The authors delve into post-COVID tourism marketing, health and wellness education and practices, ethical considerations for tourism operators, and more. A chapter also considers the challenges of sustainable supply chain management in tour operations. With contributions from experts in this emerging field, this volume is a rich resource for travel and tourism professionals, policymakers, researchers, and others. It creates a bridge between the conceptual discussions around "dark consumption" (tourism directed to places that are identified with death and suffering) and the urgency to develop empirical models that support destination marketing organizations in a rapidly changing world"-- Provided by publisher.

Identifiers: LCCN 2023020643 (print) | LCCN 2023020644 (ebook) | ISBN 9781774913505 (hardcover) | ISBN 9781774913512 (paperback) | ISBN 9781003387930 (ebook)

Subjects: LCSH: Tourism--Management. | Natural disasters--Economic aspects. | Tourism--Environmental aspects. | Dark tourism.

Classification: LCC G155.A1 P677 2023 (print) | LCC G155.A1 (ebook) | DDC 910.68--dc23/eng/20230428

LC record available at https://lccn.loc.gov/2023020643

LC ebook record available at https://lccn.loc.gov/2023020644

ISBN: 978-1-77491-350-5 (hbk)
ISBN: 978-1-77491-351-2 (pbk)
ISBN: 978-1-00338-793-0 (ebk)

About the Editors

Maximiliano E. Korstanje, PhD
Senior Researcher, Department of Economics, University of Palermo, Argentina; Editor-in-Chief, International Journal of Safety and Security in Tourism

Maximiliano E. Korstanje, PhD, is a Senior Researcher in the Department of Economics at the University of Palermo, Argentina. He is also Editor-in-Chief of the *International Journal of Safety and Security in Tourism* (Universidad de Palermo, Argentina) and Editor-in-Chief Emeritus of the *International Journal of Cyber Warfare and Terrorism* (IGI-Global US). In addition to these roles, Mr. Korstanje is a global affiliate of the Tourism Crisis Management Institute (University of Florida, USA), the Center for Ethnicity and Racism Studies (University of Leeds, UK), The Forge (University of Lancaster and University of Leeds, UK) and The International Society for Philosophers. In 2015, he was awarded a Visiting Research Fellowship at the School of Sociology and Social Policy, University of Leeds, UK, and the University of La Habanam Cuba. In 2017, he was elected as foreign faculty member of AMIT, Mexican Academy in the Study of Tourism, which is one of the most prominent institutions dedicated to tourism research in Mexico. He has vast experience in editorial projects working as an advisory member for Elsevier, Routledge, Springer, IGI Global and Cambridge Scholar Publishing. Dr. Korstanje has visited and given seminars in many important universities worldwide.

Hugues Séraphin, PhD
Senior Lecturer, Tourism, Hospitality and Events Management, Oxford Brookes Business School, United Kingdom

Hugues Séraphin, PhD, is a Senior Lecturer in Tourism, Hospitality and Events Management. He holds a PhD from the Université de Perpignan Via Domitia (France) and joined Oxford Brookes Business School in 2023. Prior to his academic career, Dr. Séraphin worked in the tourism and hospitality sector within various organizations and roles (in the Caribbean and in Europe). He was the Program Leader of the Event Management program between 2015 and 2018. This program is one of the leading programs in the country with 100% students' satisfaction (NSS, 2016). Dr. Séraphin is now the Faculty of Business and Digital Technologies Work Placement Lead. He started his career in 2003 as a French teacher in a secondary school in Birmingham. He then started to teach Travel, Tourism, and Events in 2006 when he joined a further education (FE) college in London. He possesses a broad range of international experience. This includes teaching in top Business Schools and Universities in France (EM Normandie; ESC Larochelle; ESC Pau; Université de Perpignan Via Domitia, etc.). He has received various awards for the quality of his teaching. Dr. Séraphin is also a research-active academic.

Vanessa G. B. Gowreesunkar, PhD
Associate Professor, Anant National University, Ahmedabad, Gujarat, India

Vanessa Gowreesunkar, PhD, is currently an Associate Professor at Anant National University, India. She has been teaching at various local and international institutions, such as the University of Mauritius; the University of Technology, Mauritius; Open University of Mauritius; and the Indian Institute of Travel and Tourism (India). She teaches in the fields of tourism, management, marketing, communications, German, and other areas. Dr. Gowreesunkar has published and presented a number of papers in international and local platforms and is interested in island tourism research. She is a national representative for the Organization of Tourism and Hospitality Management (OTHM), United Kingdom, and is the Chairperson and Vice President for the Scientific and Academic Research Committee (SARC) at the Allied Network for Policy Research and Actions for Sustainability.

Contents

Contributors

Claudio Bellia
Department of Agricultural, Food and Environment (Di3A) Section of Agri-Food Economics, University of Catania, Italy

Karen Cripps
Department of Responsible Management and Leadership, The University of Winchester Business School, Winchester, Hampshire, UK

Guillermo Garcia
Universidad San Pablo, Guatemala

Babu George
Alcorn State University, USA

Vanessa G. B. Gowreesunkar
Anant National University, Ahmedabad, Gujarat, India

Xavier Guegan
Department of History, The University of Winchester, Road Winchester Hampshire, UK

Maximiliano E. Korstanje
Department of Economics, University of Palermo, Buenos Aires, Argentina

Manuela Pilato
Department of Responsible Management and Leadership, The University of Winchester, Road Winchester Hampshire, UK

Hugues Séraphin
Oxford Brookes Business School, Oxford Brookes University, Oxford, UK

Simon M. Smith
Oxford Brookes Business School, Oxford Brookes University, Oxford, UK

Anca Yallop
Department of International Business Strategy and Entrepreneurship, Auckland University of Technology, Auckland, New Zealand

Maria Alejandra Zuccoli
Department of Economics, University of Palermo, Buenos Aires, Argentina

Abbreviations

ABTA	Association of British Travel Agents
AI	artificial intelligence
AoH	adult-only hotel
CBDRR	community-based disaster risk reduction
CMG	color marketing group
DMO	destination marketing organization
DR	Dominican Republic
DRA	Dale Robinson Anderson
DSC	development support communication
GDP	gross domestic product
ICTs	information and communication technologies
M&HCW	medical and healthcare waste
OECD	Organization for Economic Co-operation and Development
PAN	pensamiento automatico naturalizado
PCCD	postcolonial, post-conflict, or post-disaster
SDGs	sustainable development goals
SIDS	small island developing state
SSCM	sustainable supply chain management
ST-EP	sustainable tourism-eliminating poverty program
TTCI	travel and tourism competitiveness index
UG	undergraduate
UNWTO	United Nations World Tourism Organization
VOLM	virtual online learning material
WTTC	World Travel and Tourism Council

Introduction

"Post-Disaster and Post-Conflict Tourism: Towards a New Management Approach" was an editorial project we edited and published in 2019. In that first edition, we contemplated the global geopolitical tensions amongst nations as well as the global threats that affected the tourism industry. These dangers include terrorism, political violence, the ecological crisis, and countless natural disasters resulting from climate change. Nowadays, we are pushed to rethink a new update according to the new times in a post-COVID-19 world.

At the end of 2019, a new virus outbreak not only whipped the tourism industry as well as global trade but also shocked the world. The SARSCOv2 (COVID-19) which was originally reported in Wuhan, China disseminated in a matter of months to the four corners of the globe. Governments desperately disposed of the strictest measures to stop the virus, such as closure of airspaces, borders, and airports along with other steps such as social distancing and lockdowns.

The tourism industry, as least at we know it, withered away for several months. Scholars have debated hotly on the future of tourism just after COVID and, of course, published extensively in various leading journals (Jamal & Budke, 2020; Korstanje, 2020; Korstanje & George, 2021). However, to date, there is no recognition of the long-lasting effects of COVID-19 in the sector (Romagosa, 2020). As Baum & Hai (2020) put it, the tourism industry was the main carrier but paradoxically the main victim of COVID-19.

To review the theoretical papers published since 2020 to date is a daunting task to perform, but most certainly, they can be grouped into two great poles. On the one hand, those studies envisage COVID-19 as a unique opportunity to reverse pollution whilst changing the cultural values that led us to an ecological crisis. For these voices, tourism needs to lay the foundations for more sustainable forms of consumption (Gossling, Scott, & Hall, 2020; Ioannides & Gyimóthy, 2020; Higgins-Desbiolles, 2020). On another, other works emphasize the importance of adopting multidisciplinary methodologies to study tourism precisely in a world without tourists (Wen et al., 2020).

The present book explores a third way of filling the gap between the theory produced by Academia and the necessities of the sector. The authors who take part in this book are well-versed scholars who worked as experts in post-disaster and post-conflict tourism.

Having said this, in Chapter 1, Maximiliano E. Korstanje dissects the philosophical dichotomies beyond the term resilience. Although the term has transformed into a buzzword, it is known for its real applications. So, to say, an excess of resilience may lead people to forget the real causes of the disaster, increasing the probability it will take place again in the near future. Hence author said overtly that the notion of culture is constructed around the idea of disaster. In this vein, some practical tips to face the COVID-19 pandemic are ultimately mentioned.

In Chapter 2, Korstanje & George critically debate the ethical concerns behind post-disaster destinations just after COVID-19. Post-disaster tourism acts as a catalyst that boosts tourism consumption because of the failure of the precautionary doctrine to make it a safer destination. The term precaution sets the pace for adapting to these ever-changing days. Hugues Séraphin & Karen Cripps alert on the paradoxes of tourism development. Though tourism cultivates the grounds for prosperity and fairer wealth distribution in developing economies, under some conditions, it leads to dependency and extreme vulnerability.

Taking Haiti as a study case, the authors bring some interesting reflections on how the rebranding process potentiates or affects the tourist destination image. Chapter 1 illustrates how tourism potentiates development in postcolonial destinations. Similar remarks are found in Chapter 4, which is authored by Hugues Séraphin, Xavier Guegan & Vanessa Gowreesunkar. This chapter investigates the dynamics between concepts like history, colonial nostalgia, and tourism consumption. Of course, tourism consumption and national identity or heritage seem to be inextricably intertwined. They hold the thesis that western postcolonial branding alludes not only to an imperial past, a time of subordination and exploitation, but also denotes certain sentiment of nostalgia that prevails in the present. The chapter is based on Park's concepts speaking to us about material and immaterial remnants of the past.

In Chapter 5, Séraphin, Crips, and Smith focus on the most efficient strategies not only to grease the rails of competition but also to help the industry be back on its feet again after COVID-19. Echoing Park's argument, the authors acknowledge that heritage occupies a central position in revitalizing the identity of nations and regions as a significant vehicle toward unity and reciprocity. Manuela Pilato, Hugues Séraphin & Anca Yallop, in Chapter 6, expose an interesting case. Centered on street food in developing economies, they argue convincingly that street food conserves not only authenticity but also profitable activities for developing economies.

Chapter 7 (in the ink of Séraphin & Gowreensunkar) dissects how sports tourism boosts economies in Martinique island. This case reflects interesting inferences to be applied in a post-COVID-19 scenario. In Chapter 8, Korstanje discusses the problem of development theory, at least as it is applied in the third world. As per Harry Truman's discourse, the world was split into two: the developed and underdeveloped nations. Although a considerable amount of resources has been spent to help developing nations, no less true is that they have created a greater dependency on international corporations such as IMF or World Bank. This leads us to rethink the development theory as a dominant paradigm in the constellations of tourism. What is more important, is that the disparate access of countries to the COVID-19 vaccine reveals the material asymmetries and injustices created by global capitalism.

In Chapter 9, Séraphin and Gowreensunkar place the concept of sustainable development under the critical lens of scrutiny. In this vein, they explain that food events can be used as educational instruments at a pre-visit stage to successfully reduce the host guests' tensions or potential conflicts. The same applies to Chapter 10, where Korstanje exerts a caustic critique of the leisure inequalities between the global North and South. In a post-COVID-19 world, the chapter discusses the emergence of new forms of leisure practices as well as a new tourism behavior most certainly oriented to the other's neglect.

In Chapter 11, Korstanje and Séraphin deal with the difficulties of marketing designers and policymakers in placing the negative effects of COVID-19 in the destination image. Like other earlier chapters, this brings an interesting insight into the dichotomies between tourism development and center-periphery dependency, which is almost impossible to breach. Additionally, Zuccoli and Korstanje describe a successful study case – known as PANCOE – where students are subject to pleasurable activities such as cooking, baking pieces of bread, or even playing songs – only to name a few.

To put things straight, Chapter 8 toys with the belief that pleasure – far from creating mechanical procedures in the learning process—stimulates the cognitive attention of tourism students reducing the dropout rates whilst increasing their academic performance. Last but not least, Guillermo Garcia & Babu George – in Chapter 13– recount or relate the significant implications and intricacies of crisis management and responses before the COVID-19 pandemic. This chapter sheds light on a forward-looking perspective to understand crisis implications in a multifaceted level of analysis. The last Chapter 14, authored by Karen Cripps, offers a more than interesting diagnosis of the Sustainable Supply Chain model to be applied in the fields of tourism and hospitality. Based on a qualitative technique associated with a framework for SSCM actions, she brings an analysis of sustainable chain management in tour operations just after COVID-19.

We, the editors, want to thank the Apple Academic Press staff as well as content editor, Sandra Sickels, for their patience, generosity, and time devoted to this fascinating project. Post-Disaster and Post-Conflict tourism: an updated 2nd edition situates as a more than interesting edition that contains different chapters orientated to discuss the resilience and possibilities of adaptation of the tourism industry in a post-COVID-19 world.

REFERENCES

Baum, T., & Hai, N. T. T., (2020). Hospitality, tourism, human rights, and the impact of COVID-19. *International Journal of Contemporary Hospitality Management, 32*(7), 2397–2407.

Gössling, S., Scott, D., & Hall, C. M., (2020). Pandemics, tourism, and global change: A rapid assessment of COVID-19. *Journal of Sustainable Tourism*. DOI. https://doi.org/10.1080/09669582.2020.1758708.

Higgins-Desbiolles, F., (2020). Socializing tourism for social and ecological justice after COVID-19. *Tourism Geographies*, 1–14.

Ioannides, D., & Gyimóthy, S., (2020). The COVID-19 crisis as an opportunity for escaping the unsustainable global tourism path. *Tourism Geographies*. doi: 10.1080/14616688.2020.1763445.

Jamal, T., & Budke, C., (2020). Tourism in a world with pandemics: Local-global responsibility and action. *Journal of Tourism Futures, 6*(2), 181–186.

Korstanje, M. E., (2020). The passage from the tourist gaze to the wicked gaze: A case study on COVID-19 with special reference to Argentina. In: *International Case Studies in the Management of Disasters*. Wagon Lane, Emerald Publishing Limited.

Korstanje, M., & George, B., (2021). *The Nature and Future of Tourism: A Post COVID-19 Context.* Palm Beach, Apple Academic Press.

Romagosa, F., (2020). The COVID-19 crisis: Opportunities for sustainable and proximity tourism. *Tourism Geographies, 22*(3), 690–694.

CHAPTER 1

Tourism as a Vehicle Towards Recovery After COVID-19

MAXIMILIANO E. KORSTANJE

Department of Economics, University of Palermo, Larrea ST 1079, Buenos Aires, Argentina

ABSTRACT

The tourism industry is often vulnerable to countless risks and global threats. The COVID-19 pandemic not only ground the industry into an unparalleled halt but also has interrogated the future of the activity. This chapter critically discusses the application of resilient to make more sustainable destinations. We hold the thesis that resilience opens the doors to a paradoxical situation. On the one hand, it accelerates the recovery timeframe just after a disaster takes room. On another, it commoditizes disasters to be consumed by a global audience.

1.1 INTRODUCTION

Over the decades, scholars have agreed that tourism has shown to be a resilient industry (Hassan, 2000). As a term originally coined by Viktor Frankl, who was hosted in a concentration camp in Nazi Germany, resiliency denotes the capacity of an individual or a group to overcome extreme situations (cited in Coutu, 2002). Although widely used in the field of psychology, resiliency was recently adopted in disaster studies (Rosa, 2004). One of the limitations of the post-disaster theory rests on the

Post-Disaster and Post-Conflict Tourism (An Updated 2nd Edition), Maximiliano E. Korstanje, Hugues Seraphin, and Vanessa G. B. Gowreesunkar (Eds.)

scarcity of all-encompassing models to prevent future disasters. Precisely unpredictable disasters escape to the natural instruments of answers of community, and no less true seems to be that they vary in culture and time. Having said this, the specialized literature in built environments enthusiastically introduced the concept of resiliency as a key factor to understand not only how disasters impede human development but also those strategies which help in disaster risk reduction. Disasters are disruptive events that certainly interrogate the social background accelerating—in some conditions—substantial shifts in society. The quake of Lisbon in 1755 not only prompted the rise of a more humanist philosophy but also questioned the Theo-centric view, which later paved the way for the emergence of modern seismography (Haigh & Amaratunga, 2010).

As this argument is given, one might speculate that the reasons for and effects of disasters are varied, even if the fear of death remains the same. To wit, the notion of culture is constructed and evolved according to how successfully death is domesticated. Before the hostility of the environment, humans frequently build the contours of security, as a shelter towards the unknown. The culture is never affected by disasters rather culture is the direct creation of disaster. As Phillip Aries brilliantly acknowledged, the process of secularization, which made the advance of industrialization possible, affected negatively the influence of religion in daily life. Whether medieval peasants, who were daily subject to a context of hunger, poverty, and countless deprivations, believed in God and the afterlife, this does not happen with modern citizens. Paradoxically, though we embrace technological breakthroughs that enhance life expectancy, death has been unleashed (Aries, 1975). For Aries, death and disasters situate as founding events that alter the human cosmologies, which means the ways of conceiving our "being in this world." Equally important, the cultural entertainment industry has commoditized disasters in the forms of movies, games, documentaries, and other ideological narratives aimed at forging a morbid climate for the public. As Baudrillard (1994) puts it, the sense of "hyperreal" does not allude to the materialist conception of alienation (as Marxists originally claimed), but to the possibilities to recreate fabricated scenarios which are enrooted in the future. In this world, disasters are commoditized, transformed into reified theatricalizations, which are dubbed as "the spectacle of disaster." The most frightful aspect of disasters never seems to be their effects in the community but the possibility they repeat. As the father of modern anthropology, B. Malinowski says overtly that death plays a disruptive role in human conditions altering social ties as well as political authority. The cycles of life and death escape to human control, and of course, interrogate furtherly the culture and human institutions (Malinowski, 29).

Over recent years, modern disasters have changed the protocols of security and prevention affecting developed and underdeveloped economies worldwide. Needless to say, this was the case of the COVID-19 pandemic, and the SARSCOv2 virus outbreak originally reported in Wuhan, China. Since the first edition of the present book, many things have happened. The book, which was originally published in 2018, lacks chapters and content discussing the long-lasting effects of COVID-19. What is more important, the pandemic not only grinds the tourism industry but also the global trade to a halt. Governments' desperately disposed of restrictive measures to contain the virus dissemination which included the border and airspace closure without mentioning the imposition of strict lockdowns and social distancing. All these measures changed forever our travel behavior. In the mid of this mayhem, some questions arise: what is the role of post-disaster tourism and management in dealing with the pandemic? In what way would this pandemic be avoided? What are the safety and security protocols in tourism to deal with virus outbreaks? What are the lessons left by COVID-19 and other potential pandemics?

Some studies have attempted to respond to these above-noted questions in the past. For example, interesting discussions are given on the fact that disasters affect in short or long-turn to the organic image of tourist destinations can be found in Ritchie (2004); Huang, Tseng, & Petrick (2008); and Hall (2010). Hugues Séraphin offers a post-disaster marketing model oriented to organize post-disaster recovery facet just after a disaster takes hit (Séraphin, Butcher, & Korstanje, 2017; Séraphin et al., 2016).

As the previous backdrop, one of the more remarkable aspects of COVID-19 respecting other disasters is its global long-lasting consequences on the tourism industry. In addition, Academia provided different viewpoints and diagnoses about how to deal and how to do with the COVID-19 pandemic. Whilst some voices have applauded the possibility to re-found the tourism industry towards more sustainable forms of consumption (Gossling, Scott, & Hall, 2020; Higgins-Desbiolles, 2020; Romagosa, 2020), others alluded to the "end of the tourism industry" as least as we know it (Korstanje & George, 2022). It is important to mention that the problem of COVID-19 needs a further discussion on how resilient the tourism industry is. The tourism industry is paradoxically the main carrier and victim of COVID-19 (Baum & Hai, 2020; Korstanje & George, 2022). Some of the sections in this second edition have been updated or changed to offer readers a full snapshot of post-disaster studies, their advantages and disadvantages—to help the industry to keep back on its feet again.

1.2 A SHORT INTRODUCTION TO RESILIENCE PARADOXES

The present chapter discusses critically the role played by the tourism industry as an instrument of resiliency which enhances not only the affected social ties but contributes directly to the community recovery facet. It is difficult to resist the impression that even if post-disaster marketing assists policymakers in the recovery stage, no less true seems to be that a paradoxical situation surfaces. Disasters are often commoditized and exchanged as esthetic consumable goods to be gazed at by tourists. When this happens, the real reasons for disasters are distorted or forgotten in the threshold of time. It is important to note that post-disaster tourism – far from reducing the structural factor that facilitated the event—paves the way for a spectacle that captivates global audiences. This is the exact example of Ground-zero who has become in highly visited tourist attraction at the time. Scholars and policymakers cast some doubts revolving around to what extent tourism opens the doors toward recovery or relates to the fact the event will repeat itself in the future. Here is where the problem lies: when the lessons of disasters are not learned or remain ignored, the possibilities of the structural factors facilitate a new disrupting event turns higher. The commoditization of disasters is functional to the disaster itself. The ethical dilemma is how we could reduce future disasters if they are the main commodity to exchange in a society fascinated by these types of spectacles.

1.3 WHAT IS A DISASTER?

Disasters are normally seen as external catastrophic events which place society in jeopardy. However, as sociologists of disasters agree, disasters should be understood as cultural social constructions previously imagined and negotiated in much deeper social conventions (Piran, 2004). Enrico

Quarantelli, a well-renowned expert in disaster studies – and Distinguished Professor from Delaware University – defines disasters as social construes independent of human actions that usually affect the normal life of a community or the normal lines of perception. What can be horrendous for a community seems not to be for others. Hence, there are no concrete numbers of victims and psychical losses that certainly mark what a disaster is. Disasters not only threaten society but also engage with some specific collective attitudes and reactions which suspend the legitimacy of the law. To set a clear example, whilst kicking a refreshment machine is seriously sanctioned in daily life, this action may help various lives in the mid of an earthquake. At the same time, looting is rarely infrequent in a post-disaster context. Rather, people tend to cooperate in dangerous situations assisting others with shelter and food (Quarantelli, 2005). Even Quarantelli goes beyond further, arguing convincingly that how risky a situation is perceived, and ethnicities are inextricably intertwined. Whilst some ethnic minorities are less receptive to risk perception, others are prone to exaggerate the emergency messages disseminated by the mass media (Quarantelli, 1995, 1996, 2005). In this vein, some critical philosophers like Paul Virilio (2010); and Jean Pierre Dupuy (1995) alert that disasters are often commoditized, exchanged or sold according to the demand of wider audiences in Western nations. This invariably creates a culture of fear, which is based on the precautionary doctrine. The precautionary doctrine punctures the urgency of locating and eradicating global risks through rational planning. Sometimes this does not happen simply because we are educated to inflate some minor risks whilst other major dangers are culturally overlooked (Sunstein, 2005). Typically, some scholars are accustomed to thinking disaster ignites panic flights which are devastating for society. In his book, El Panico (The Panic) Jean Marie Dupuy resumes the earlier works of classic scholars such as Locke, Rousseau, and Hobbes who were originally interested in understanding how psychological fear works. Both – but from different angles – reach a consensus that fear – far from being destructive—keeps society united. Whilst Hobbes used the metaphor of the War of all against War to legitimate his figure of Leviathan Locke alluded to the fear of Civil Laws to acknowledge that fear is culturally enrooted in the ideological core of society. Dupuy cites the myth of Pan, an ancient Greek mythological deity who harassed and killed travelers when they are lost. The term panic, doubtless, comes from the figure of Pan who remained in silence during the entire trip planning the moment to attack foreign travelers. The same applies to the sentiment of panic which remains occulted over years waiting to see the light of publicity. Based on the legacy of L. Crocq, Dupuy notes that the panic should be understood as a collective and extreme fear which simultaneously connects people to an archaic state of anomie where laws are temporarily suspended. At the time, French cultural theory suggests that panic obeys to mimicry, or at the best, to the capacity of humans to imitate others, Americans preferably accept the subject never returns to the natural state. The panic, unless otherwise resolved, leads to extreme individualism, which escapes to critical thinking (Dupuy, 1999). Dupuy subverts the notion of narcissism, at least as it was originally underpinned by S. Freud and his followers. For Dupuy, Freud never approached the problem of Pseudo-narcissism in his writings. At a closer look, as Dupuy scrutinizes, the pseudo-narcissism is socially enrooted in the core of the marketplace endorsing the charisma over political authorities. The marketplace should be seen as the source of power in modern societies. The power does not emanate from the law, and the power emanates from the pseudo-narcissism that gives political leaders to be excepted by the law. In this way, lay citizens profess love for the "Other" whilst imitating the love that their leaders have for themselves. The market, in sum, contains an ancient fear which is ultimately liberated once the disaster takes a hit (Dupuy, 1999). In a seminal book entitled "*The University of Disaster*," Paul Virilio (2010) sheds light on the problem of modern fear and its expansion through a global world. Modern capitalism expanded successfully through the articulation of technological breakthroughs that accelerated the production cycles and frames. The world has been transformed into a fictionalized landscape where "the Other" is neglected, if not eradicated. Although global capitalism gives to foreign tourists a great multiplicity of landscapes to be gazed at, they had no direct contact with locals. In ancient times, the wall marked the inside from the outside, protecting the city from strangers or intruders. Today, the advances of technologies in a high-mobile cosmopolitan society blur the borders between here and there, homogenizing and standardizing the urban life in cities. At a closer look, the neighbor has become a stranger for us. Last but not least, the net of experts seems to be more interested in finding and eliminating the risks of businesses than in protecting the environment or people. The philosophy of science is today determined by the logic of digital screens. The simulation of the future that characterizes the digital world has replaced daily life. Of course, this raises an interesting question about the role of experts in this process. However, the theme does not rest there, as an ally of the market, science devotes its resources to implementing complex algorithms intended to forecast the future. Risks should be calculated, found, and eradicated to prevent the disaster. The paradox, doubtless, lies in the fact that whilst humans intervene to achieve a zero-risk society, they lay the foundations for a new catastrophic event (Virilio, 2010). For Virilio, disasters are nothing other than mere spectacles widely packaged and disseminated by the mass media, a type of new media entertainment that captivates the attention of audiences globally.

1.4 MODERN DISASTERS AND MEDIA ENTERTAINMENT

In earlier sections, we have discussed to what extent modern disasters are fabricated and packaged as a modern form of entertainment in Western audiences. This section goes in the opposite direction. Resiliency plays a leading role in the post-recovery stage. Having said this, Professor Rajib Shaw (2012) describes with accuracy the intersection of disasters, environment, and community response. He introduces readers to the development of community-based disaster risk reduction (CBDRR) which was a term widely used in the specialized literature and studied throughout the world. Quite aside from this popularity, Shaw explains few studies have certainly focused on practical cases examining not only the history of CBDRR but also delving into the necessary steps for enhancing community participation in the process of mitigation and recovery. The university and academicians are more than important in giving communities the necessary knowledge to expand the current understanding for improving the evacuation in contexts of emergencies. Performatively, the spirit of corporations that employ residents may give certain stability to accelerate the recovery time frame. Altogether with Governments, NGOs, and Corporations financial profits are not incompatible with the ethical issues and the stakeholders' interest in post-disaster contexts. Diverse examples prove that international organizations should not ignore that capital can improve community resiliency. Secondly, the introduction of business corporations in the process of reconstruction can generate a gap between survivors and State. First and foremost, in the Third world, what residents expect from their governments is fulfilled by the market and international corporations that re-cycle obliterated terrains at lower costs and sell later for the exorbitant sum of money. As ground-zero in NY became a tourist attraction, post-disaster lands draw the attention of international investors who to some extent help residents to accelerate the recovery but at the same time, they replicate the same reasons that made the disaster possible. Furthermore, these results in a process of gentrification using which some actors are obliged to abandon their homes to be re-located in more dangerous and peripheral sites. The decline or lack of intervention of States in these situations may have negative aftermaths for the community, even when CBDRM is applied.

American psychologist Kai Erikson (1994) calls attention to the need of deciphering the much deep interconnection of disaster and cultural consumption. He narrates the case of Ojibwa, an Indian reservation situated nearly to Grassy Narrows in Canada. On this site, the abuse and addiction to alcohol are wreaking havoc, provoking a high prevalence of crimes and murders. Erikson toys with the belief disasters whilst imagined are not hitting in reality. There are a lot of humanitarian disasters which remain obscured in the social imaginary. In Grassy Narrows, the alcohol and drug abuse are gradually effacing the social ties and social scaffolding. Like natural disasters, dwellers are often subject to the climate of depersonalization, violence, and social fragmentation. In the winter of 1970, Erikson witnessed as the Wabiggon's River has been gradually contaminated with mercury. The fact was that the people of Ojibwa ignored the fact that they were exposed to this poison for more than one decade. The effects of this were not observable in the short run; clinical symptoms were associated with numbness of the mouth, problems in the brain, clumsiness in walking, memory loss, impaired visions and finally depression. Undoubtedly, mercury caused havoc in the population of Grassy Narrows. However, mercury was not the only threat this community faced. White's colonization pushed natives to change their ancestral customs and form of tribe organization. This progressive disintegration of clans in combination with difficulties to track down and hunt animals made survival problematic for Grassy Narrows's natives.

Disasters have historically captivated the imagination of public opinion and global audiences (Quarantelli, 1985). This happens because few people experience in their life a real disaster. As a phenomenon that opens our imagination towards other horizons, disaster-consuming is a fertile ground for the popular culture and media industry. One of the authoritative voices in Disaster-movies as a novel genre is Stephen Keane. Almost a decade ago, he published a book which is entitled "*Disaster Movies: The Cinema of Catastrophe*." In this seminal text, Keane (2006) alludes to disaster films as a neglected theme in cinema studies. This particularly obscures the understanding laypeople have in this genre. Although disaster movies adopted different shapes and lessons over time, no less true is that after the 90s decade, a new purer and hybrid style surfaced. Next to the end of the millennium – from tornadoes to aliens – disaster movies concentrated a lot of diverse anxieties, which were commoditized into a fashionable product: the apocalyptic landscape. As a form of domesticating death and destruction, disaster movies bring the ideological reading behind more practical and industrial reasoning into the foreground. To put the same in other terms, Hollywood portrays disasters that happen one day, but are constantly replicated over years. With the end of repeating the same thrilling product, media maximizes costs and times as an essential sharable commodity to wider audiences. In consequence, we feel a pleasurable experience that is previously determined by the spectacle of disaster. Glenn Kay & Michael Rose (2006) argue convincingly that disaster movies have been so popular because of a morbid fascination coupled with the idea the world someday will end. This creates some states of anxieties these types of movies mitigate. This begs the question to what extent the essence of these movies may be extrapolated to tourism consumption.

In the book "*The Rise of Thana Capitalism and Tourism*," we ignited a hot debate around the concept of Thana-Tourism, which is defined as a new postmodern trend where tourists visit spaces of mass death, loss, pain, and mourning. However, per my appreciation, this does not limit to tourism, the obsession for gazing at other's death is applied to all popular culture. The society of risk, as it was imagined by postmodern sociologists set the pace to a new stage of capitalism: Thana-Capitalism. The introduction of secularization, which prompted the decline of religion, derived from two observable issues. On the one hand, people are afraid of dying, simply because they believe in nothing. Death is often neglected or as in the case of disaster-related films sublimated into an esthetic of morbid consumption. On another, disasters remind us how special we are. After all, the gods have touched us and protected us from total obliteration. This recreates a narcissist spirit which – unless regulated – may lead towards closed and chauvinist expressions. In the days of Thana-Capitalism, we individually consume the others' death to reinforce a sentiment of supremacy as "survivors" who are still in the trace (Korstanje, 2016). This aspect of my account was applauded but criticized by R. Tzanelli in a professional review that I want to share with readers. Tzanelli enthusiastically acknowledges that death is being commoditized as never before – mediating between citizens and their institutions-, but she alerts on the risks of extrapolating eclectic considerations revolving around the idea of a sadist society, as she reads I mistakenly did. Tourists are not moved by only one motive; they are motivated by countless reasons and motives. This suggests that even in the act of consuming morbid landscapes, many stereotypes and expectancies converge. Secondly, Thana-Tourism represents an act of empathy with disgrace, which helps to understand our finitude (Tzanelli, 2017; Korstanje, 2016). Though her posture sounds illustrative and interesting, we replied that one of the most salient features of Thana-Capitalism was based on the lack of empathy for others, which adjoined to pathological solipsism is legitimated by "the war of all against all" (citing Hobbes).

1.5 THE EMOTIONALITY OF THE TRAUMA

In the thanaptopic society, any disaster and space of trauma are ultimately packaged as a tourist attraction. The risk society, at the best as it was imagined by postmodern sociologists, has set the pace to a new stage of production where the "Other's pain" is commoditized to revitalize the daily social frustration. In a nutshell, the other's death acts as a catalyst instrument to endorse the global audience a quota of adrenaline, happiness, and pleasure. As discussed, these trends are not limited to dark tourism, they can be very well found in different spheres of the media entertainment industry encompassing novels, TV programs, Video Games and so forth (Korstanje, 2016). The postmodern self only maximizes his pleasure through the other's pain. In the ecstasy of the spectacle, lay-citizen temporarily forgets the frustration and the oppressive working conditions where he moves. At a first glimpse, capitalism centers on two main pillars, the need of being special and the need to be protected (Korstanje, 2016). In consonance with this, Tiya Miles describes brilliantly how life in the Imagined (Black) South in the US engages with the macabre taste for the darkness many tourists show. Per her viewpoint, the ideological construction of the darkness is simply an absence that is filled by capitalist exploitation. Per her words as ethnographers, dark consumption is a continuation of southern slavery, so to speak that: "Nearly a decade before when I had begun my research on southern slavery, black history has been all but ignored at the historic home and plantation sites. The people who have mattered in these tours were slaveholding high society families, not their chattel slaves. African American bondsmen and bondswomen had been transformed into virtual ghosts, absent, and yet eerily present in historical tours as invisible laboring bodies that made their owners' fortune shine" (Preface, p. xxi).

The dark tours were organized to show the landscapes of the haunted south, where the alterity was historically enslaved, and subject to a climate of oppression and cruelty. The voodoo practices, many of them strongly associated with slaves seems to be the symbolic touchstone of these tours. Visitors not only are interrogated by the tragedy of slavery, but it is accompanied by personal histories and biographies characterized by human sacrifices and black magic. Miles' main thesis holds that the axiom is that dark tourism is sociologically adjusted to a heritage, which is politically manipulated to serve as a spectacle fabricated for the white lords and widely consumed by international visitors. In so doing, the real history of oppression and violence is subverted whilst the Alterity is silenced. The pleasure of consuming traumatic experiences is given by the need of consuming a distorted (tailored) story to legitimate the status quo. The more social fragmentation and climate of depersonalization, the more importance dark tourism takes. As R. Tzanelli puts it (2016), the cinematic representations of dark tourism are externally designed to be enjoyed by first-class tourists, many of them coming from sophisticated urban cities. Though human suffering is the main attraction in the peripheral south, there is no genuine remorse for the actions and responsibilities of their respective states during slavery and colonial rule. In sharp opposition, Dorina Buda argues that dark should be framed as a psychological drive enrooted in the Thanatos. She dissects dark tourism escaping from any eclectic interpretation. Buda re-assumes Freud's contributions to social geography studies. What she claims, professional researchers have avoided adopting psychoanalytical advances considering them as pseudo-science. This is a great flaw simply because they remain unaware of the real factors behind our decisions. Psychoanalysis offers a fertile ground to understand the complexity of dark tourism. At a closer look, dark consumption reflects the death drive as a powerful force that is placed in opposition to life – Eros. Her basis on a study case conducted in Palestine-Israel calls attention to the need of introducing a new term, emotional geography. Negative emotions such as fear or even shock is automatically repelled towards the emotional contours of self. These special borders not only are checkpoints or borderlands where the sense of security is debated but also where fear is condensed. To put the same in other terms, the visit to effaced or conflict-zones revives not only old long-dormant phantoms but also it associates to a traumatic experience the visitor wants to rememorize, at least unconsciously. The decision to travel to dark sites should be understood as an attempt to rememoration of a much deeper trauma. This occurs because our decisions dominate the death drive. To sum up, she speaks to us of the geographies of death to signify the interconnection of the self to a traumatic space. Similar reflections can be found in Maria Tumarkin's texts. In a broader sense, Tumarkin (2013) coins the term traumascape to denote the abilities of the self in embodying traumatic spaces which very well replicate and resonate over time. She goes openly on to say, "I would like to suggest that Port Arthur is a traumascape, which I define as a time place materially and discursively bound by traumatic repetitions. This statement, which will be unpacked in detail in the following section, has very real and tangible consequences for our understanding of the legacy of Port Arthur and may shed some light on the explosive and deeply ambivalent nature of the massacre's aftermath. I am, in essence, saying that Port Arthur is a different kind of place to other sites of major tragedies, different to say, Newcastle. As a different kind of place, it demands different kinds of ideas about survival, trauma, memory, and the effect of trauma on the landscape and the local community. These different kinds of ideas have a direct bearing on public policy measures and governmental involvement in the short and long-term recovery process" (Tumarkin, 2001, p. 199).

With the benefits of hindsight, one might speculate that this discussion led us to think that probably tourism should be comprehended as a mechanism of resilience. Is tourism part of the problem or the solution?

Many studies have evinced how tourism is a commercial activity sensitive to global risks and dangers. Tourism, which evolves in multiple service sectors, manifests the impossibility of the buyers (tourists) to enjoy the product earlier than consumed. Of course, this means that tourism has an evident retractile nature (Ritzer & Jurgenson, 2010; Ritzer, 2015). The turn of a new millennium brought many unseen risks like terrorism, lethal virus outbreaks (like COVID-19), natural disasters and one of the major stock and market crises of capitalism in its history. Unfortunately, tourist destinations were covered by the veil of uncertainty, and many policymakers claimed that new forms and programs of protection should be implemented. The present section comments on some of the advances and limitations of applied research in understanding to what extent global risks affect tourist destinations.

One of the original works in this direction should be traced to Castaño (2005) who gathers statistical information respecting the recovery of tourism in post-disaster scenarios. Per his viewpoint, tourist destinations not only recover after a disaster or a terrorist attack but also surpass the flux of visitors and profits to upper levels earlier to the disrupting event. This led to Korstanje & Ivanov (2012) to confirm that tourism acts as a mechanism of resilience within society by the revitalization of material losses but articulating a narrative as well for next generations. In consonance with this, Duan Biggs (2011) sets forward an interesting model to mitigate the effects of global warming in tourism. He departs from the premise that human capital is a key factor in the strengthening of enterprise resilience. Besides, Korstanje & Clayton (2012) questioned the effects of terrorism on tourism industries. Instead of defining terrorism as a threat, authors polemically argue that terrorism revitalizes the attractiveness to tourist sites through the articulation of new morbid consumption. This hypothesis is empirically validated by other countless studies such as Saha & Gap (2014); Raine (2013); and Yan et al. (2016). Whatever the case may be, today the paradigm of disaster studies is being incorporated into the planning literature, which suggests that resilience is the centerpiece of sustainability. What is more important, tourism development plays a vital role in providing economic and financial aid for communities as well as pertinent support to enhance local resiliency (Kato, 2017). Miller, Gonzalez, and Hutter published a pungent paper discussing the conceptual limitations and strongholds of dark tourism to be applied as an instrument of recovery in case of disasters. Like the case of Phoenix tourism, tourism allows the rebirth of places after disasters and traumatic events. The tourism industry revitalizes the affected image according to the new lens which involves a rebranding process. Transforming the mourning into an allegorical spectacle is not an unethical task, but the opposite, as these authors showed. Dark tourism and post-disaster marketing help locals to convert the negative into positive assets which reinforces the human capital.

Nonetheless, some critical authors emphasize the negative effects of post-disaster programs such as real-estate speculation, inflation, inter-class or ethnic conflict and problems associated with a gentrification process. In this context, Pezzullo (2009) analyzes the case of New Orleans, a city that has been effaced by the Katrina hurricane in 2005. Katrina was one of the worst natural disasters the US faced in history. For days, people were dramatically stranded without help, food, shelter, and means of mobilities. This sad event almost killed 1,577 citizens causing total losses of US $81 billion. Despite this state of great mourning, in September of the same year, Gray Line announced the creation of Katrina Tours which constituted a new attempt to attract tourists offering the devastation left by Katrina as the main attraction. Pezzullo starts to discuss consuming disasters not only may be an opportunity to create catharsis but also opens the doors for a new form of education. Since disaster has not been predicted, at least, tourism would assist survivors in rebuilding the city. Katrina tours invited tourists to preserve the place as a national icon, as Pezzullo adheres, in the same way, the previous relations of power changed. These tours rearticulated an essentialist national being because of the New Orleans rebirth, drawing new geography and national negotiation among stakeholders. In a similar structuring of emotions than 9/11, Katrina tours appealed to the reciprocity and solidarity only disasters engage but leaving the material asymmetries in the pre-Katrina conjuncture. This poses a serious ethical doubt respecting to what extent dark tourism helps or hides the conditions that ultimately generated the state of emergency or the disaster.

1.6 THE SPECTACLE OF COVID-19

The first cases of COVID-19 were supposedly reported in Wuhan, China. However, Health authorities and scientists do not agree to precise the origin of the virus nor the conditions of its propagation. What seems to be more important, Chinese tourists, as well as European businessmen, were the first carriers who involuntarily disseminated the virus (Jones & Comfort, 2020; Rogerson & Rogerson, 2020). The first cases arrived in Italy and Spain as a result of the high trade exchange between China and Europe (Yee, 2020). As discussed, the tourism industry was the main carrier and victim of the new COVID-19 disease. To date (on 6 October of 2021), authorities recognize more than 4 million deaths and 200 million infected worldwide. The most affected countries were the US, India, Brazil, the UK, Russia, and Mexico – only to name a few. The authorities and governments disposed of restrictive measures to contain the infection, but in some cases, they were later adopted. These measures included the borders and airspace closure, associated with the imposition of strictest lockdowns and social distancing. Thousands of tourists were stranded abroad, without medicine, assistance, shelter, or food (Prayag, 2020; Korstanje, 2021). Governments commanded their flag carriers to repatriate their citizens stranded abroad.

These dramatic events were aggravated by inter-state geopolitical tensions, accompanied by chauvinist and racist expressions against Asian tourists (Wessler & Talarico, 2021; Yu et al., 2021; Mostafanezhad, Cheer, & Sin, 2020). As Korstanje & George (2022) observed, COVID-19 re-framed not only the human relationships introducing fear as a mediator between citizens and their authorities but also the global order towards a new feudalized (if not fractured) world. The tourists who historically were envisaged as ambassadors of developed and civilized countries are now suspected and demonized as carriers of a lethal virus. Furthermore, the pandemic opened the doors towards a new morbid spectacle where "the Other's death" was the main commodity to exchange. The unilateral restrictive measures plugged lay citizens to the macabre show of COVID-19 24/7. Every day, health authorities reported the number of deaths, infected people and recovered patients. The audience remained captive of this moment for months. COVID-19, unlike another virus, affected particularly elderly people causing severe symptoms such as muscle pain, cough even breathing difficulties. Other patients developed mild symptoms such as fever, nasal congestion or loss of smell and taste. One of the aspects that terrified the audience links to the fact nobody knows really who would develop severe breathing problems or not. COVID-19 was like playing with a great lottery. As stated, the restrictive measures varied on country and culture, but even under the same national jurisdiction. Some countries experienced a type of much deep fragmentation (among provinces, counties, or neighbors). Let's explain to readers that even if WHO fixed an agenda of protocols and counter-infection measures, each country adopted different paths and steps to contain the pandemic. Whilst some nations banned foreign tourism, others moved through the strictest quarantines. Chauvinist and separatist – long-dormant – discourses ultimately surfaced. The inter-ethnic conflicts not only revived but also the old geopolitical tensions among some nations. Last but not least, COVID-19 offered a show where tourists were the bad guys. Newspapers and TV newscasts reported that many foreign tourists violated the quarantine at hotels, putting public health at serious risk. In some cases, tourism staff refused to give attendance to foreign tourists at some international tourist destinations. News showing tourists sleeping at airports or stranded abroad was normal during the COVID-19 context. Whether John Urry coined the term "tourist gaze" to denote the dominance of a cultural matrix that indicates what can be gazed at or not, we dubbed the term "wicked gaze" to refer to a new state of capitalism where the foreign tourist is an undesired guest. The point represents very well the end of hospitality and globalization as we know it (Korstanje & George, 2022). The lessons given by the pandemic are twofold. On the one hand, it is very hard to precisely why the precautionary doctrine and the forecasting instruments failed to predict or mitigate the COVID-19 aftermaths. Even though some voices cautioned on the possibilities a new pandemic placed the planet in jeopardy, health authorities ignored the alert. Some Chinese activists were systematically silenced by the central administration. The alert to the Western countries was not correctly deciphered. China delivered experts to Western nations to help in struggling against the virus, but their pieces of advice were unfortunately underestimated. On the other hand, a rivalry between China and the US impeded coordinated efforts against the pandemic. With ebbs and flows, each country adopted its proper measures and steps to contain the pandemic. Everything that has been written about pandemics failed to deal with COVID-19. Needless to say, the pandemic created a new fractured world where the "Other" is suspected and feared. The national borders are in constant change and negotiated by local actors. As Romagosa (2020) noted, the future of tourism remains uncertain and open in the middle of this disaster.

1.7 CONCLUSION

The present chapter updates and discusses the future of the tourism industry in a post-COVID-19 world. In so doing, we stress the importance of resilience to find new innovative solutions for accelerating industry recovery. Post-disaster tourism offers a promising model to understand the next steps to follow. Unlike the precautionary principle which emphasizes the urgency of pre-emption to mitigate risks, post-disaster tourism operates underway the facts are happening. This chapter introduces readers to the discussion that leads to creating a gap between practice and theory. Over the recent years, the tourism industry has been subject to countless global dangers oscillating from natural disasters and outbreaks to terrorism and political violence. Of course, the effects affected more to underdeveloped economies than developed ones. The Global South often dependent on global tourism showed no fewer problems to overcome the global challenges such as the effects of climate change or the ecological crisis. Having said this, the emergence of a new theoretical paradigm that helps policymakers to deal with these new threats appears to be necessary. This book centers on the old stereotypes and prejudices historically enrooted in Western colonialism as well as in the material asymmetries of the Global South. We have exerted a radical criticism on dark tourism consumption, but it is safe to say that dark consumption and post-disaster marketing are today valuable resources that help devastated communities in the recovery facet.

KEYWORDS

- **commodity**
- **COVID-19**
- **disasters**
- **global risks**
- **resilience**
- **tourism**

REFERENCES

Ariès, P., (1975). *Western Attitudes Toward Death: From the Middle Ages to the Present* (Vol. 3). Baltimore, MA: John Hopkins University Press.

Baudrillard, J., (1994). *Simulacra and Simulation.* Michigan, MI: University of Michigan Press.

Baum, T., & Hai, N. T. T., (2020). Hospitality, tourism, human rights and the impact of COVID-19. *International Journal of Contemporary Hospitality Management, 32*(7), 2397–2407.

Biggs, D., (2011). Understanding resilience in a vulnerable industry: The case of reef tourism in Australia. *Ecology and Society, 16*(1), 1–10.

Buda, D., (2015). *Affective Tourism, Dark Routes in Conflict.* Abingdon: Routledge.

Castaño, J. M., (2005). *Psicología Social de los Viajes y del Turismo [Sociopsychology of Travels and Tourism].* Madrid: Thomson.

Coutu, D. L., (2002). How resilience works. *Harvard Business Review, 80*(5), 46–56.

Dupuy, J. P., (1999). *El Pánico [Panic].* Barcelona: Gedisa.

Eriksson, K., (1994). *A New Species of Troubles, Explorations in Disasters, Trauma and Community.* New York, NY: WW Norton and Company.

Gössling, S., Scott, D., & Hall, C. M., (2020). Pandemics, tourism and global change: A rapid assessment of COVID-19. *Journal of Sustainable Tourism, 29*(1), 1–20.

Haigh, R., & Amaratunga, D., (2010). An integrative review of the built environment discipline's role in the development of society's resilience to disasters. *International Journal of Disaster Resilience in the Built Environment, 1*(1), 11–24.

Hall, C. M., (2010). Crisis events in tourism: Subjects of crisis in tourism. *Current Issues in Tourism, 13*(5), 401–417.

Hassan, S. S., (2000). Determinants of market competitiveness in the environmentally sustainable tourism industry. *Journal of Travel Research, 38*(3), 239–245.

Higgins-Desbiolles, F., (2020). Socializing tourism for social and ecological justice after COVID-19. *Tourism Geographies, 22*(3), 610–623.

Huang, Y. C., Tseng, Y. P., & Petrick, J. F., (2008). Crisis management planning to restore tourism after disasters: A case study from Taiwan. *Journal of Travel & Tourism Marketing, 23*(2–4), 203–221.

Jones, P., & Comfort, D., (2020). The COVID-19 crisis, tourism and sustainable development. *Athens Journal of Tourism, 7*(2), 75–86.

Kato, K. (2018). Debating sustainability in tourism development: Resilience, traditional knowledge and community: A post-disaster perspective. *Tourism Planning & Development, 15*(1), 55–67.

Kay, G., & Rose, M., (2006). *Disaster Movies.* Chicago, IL: Chicago Review Press.

Keane, S., (2006). *Disaster Movies: The Cinema of Catastrophe.* London: Wallflower.

Korstanje, M. E., & Clayton, A., (2012). Tourism and terrorism: Conflicts and commonalities. *Worldwide Hospitality and Tourism Themes, 4*(1), 8–25.

Korstanje, M. E., & George, B., (2022). *The Nature and Future of Tourism Research: A Post COVID19 Context.* Palm Bay, FL: Apple Academic Press.

Korstanje, M. E., & Ivanov, S. H., (2012). Tourism as a form of new psychological resilience: The inception of dark tourism. *Cultur, 6*(4), 56–71.

Korstanje, M. E., (2021). Habeas corpus: Argentinean tourists stranded. *Tourism and Hospitality, 2*(4), 327–331.

Korstanje, M., (2016). *The Rise of Thana Capitalism and Tourism.* Abingdon: Routledge.

Malinowski, B., (1954). *Magic, Science and Religion.* Boston, MA: Beacon Press.

Miller, D. S., Gonzalez, C., & Hutter, M., (2017). Phoenix tourism within dark tourism: Rebirth, rebuilding and rebranding of tourist destinations following disasters. *Worldwide Hospitality and Tourism Themes, 9*(2), 196–215.

Milles, T., (2015). *Tales from the Haunted South.* Chappel Hill, NC: The University of North Carolina Press.

Mostafanezhad, M., Cheer, J. M., & Sin, H. L., (2020). Geopolitical anxieties of tourism: (Im) mobilities of the COVID-19 pandemic. *Dialogues in Human Geography, 10*(2), 182–186.

Pezzullo, P. C., (2009). This is the only tour that sells: Tourism, disaster, and national identity in New Orleans. *Journal of Tourism and Cultural Change, 7*(2), 99–114.

Piran, P., (2004). The role of disasters sociology in crisis management. *Social Welfare Quarterly, 3*(11), 13–48.

Prayag, G., (2020). Time for reset? COVID-19 and tourism resilience. *Tourism Review International, 24*(2, 3), 179–184.

Quarantelli, E. L., (1985). Realities and mythologies in disaster films. *Communications, 11*(1), 31–44.

Quarantelli, E. L., (1995). Patterns of sheltering and housing in US disasters. *Disaster Prevention and Management: An International Journal, 4*(3), 43–53.

Quarantelli, E. L., (1996). The future is not the past repeated: Projecting disasters in the 21st century from current trends. *Journal of Contingencies and Crisis Management, 4*(4), 228–240.

Quarantelli, E. L., (2005). *What is a Disaster? A Dozen Perspectives on the Question.* Abingdon: Routledge.

Raine, R., (2013). A dark tourist spectrum. *International Journal of Culture, Tourism and Hospitality Research, 7*(3), 242–256.

Ritchie, B. W., (2004). Chaos, crises and disasters: A strategic approach to crisis management in the tourism industry. *Tourism Management, 25*(6), 669–683.

Ritzer, G., & Jurgenson, N., (2010). Production, consumption, prosumption: The nature of capitalism in the age of the digital 'prosumer.' *Journal of Consumer Culture, 10*(1), 13–36.

Ritzer, G., (2015). Automating prosumption: The decline of the prosumer and the rise of the prosuming machines. *Journal of Consumer Culture, 15*(3), 407–424.

Rogerson, C. M., & Rogerson, J. M., (2020). COVID-19 and tourism spaces of vulnerability in South Africa. *African Journal of Hospitality, Tourism and Leisure, 9*(4), 382–401.

Romagosa, F., (2020). The COVID-19 crisis: Opportunities for sustainable and proximity tourism. *Tourism Geographies, 22*(3), 690–694.

Rose, A., (2004). Defining and measuring economic resilience to disasters. *Disaster Prevention and Management: An International Journal, 13*(4), 307–314.

Saha, S., & Yap, G., (2014). The moderation effects of political instability and terrorism on tourism development: A cross-country panel analysis. *Journal of Travel Research, 53*(4), 509–521.

Séraphin, H., Ambaye, M., Gowreesunkar, V., & Bonnardel, V., (2016). A marketing research tool for destination marketing organizations' logo design. *Journal of Business Research, 69*(11), 5022–5027.

Séraphin, H., Butcher, J., & Korstanje, M., (2017). Challenging the negative images of Haiti at a pre-visit stage using visual online learning materials. *Journal of Policy Research in Tourism, Leisure and Events, 9*(2), 169–181.

Shaw, R., (2012). *Community-Based Disaster Risk Reduction.* Bingley: Emerald Group Publishing.

Sunstein, C. R., (2005). *Laws of Fear: Beyond the Precautionary Principle* (Vol. 6). Cambridge, MA: Cambridge University Press.

Tumarkin, M., (2001). Wishing you weren't here: Thinking about trauma, place and the port Arthur massacre. *Journal of Australian Studies, 25*(67), 196–205.

Tumarkin, M., (2013). *Traumascapes: The Power and Fate of Places Transformed by Tragedy*. Melbourne: Melbourne University Publishing.

Tzanelli, R., (2016). *Thanatourism and Cinematic Representation of Risk*. Abingdon: Routledge.

Tzanelli, R., (2017). Bios and Thanatos in leisure: A critical review of Maximiliano Korstanje's the rise of Thana-capitalism and tourism. *European Journal of Tourism Research, 17*(1), 269–272.

Virilio, P., (2010). *The University of Disaster*. Oxford: Polity Press.

Wassler, P., & Talarico, C., (2021). Sociocultural impacts of COVID-19: A social representations perspective. *Tourism Management Perspectives, 38*, 100813.

Yan, Bing-Jin., Zhang, J., Zhang, H. L., Lu, S. J., & Guo, Y. R., (2016). Investigating the motivation–experience relationship in a dark tourism space: A case study of the Beichuan earthquake relics, China. *Tourism Management, 53*, 108–121.

Yeo, L. H., (2020). Impact of COVID-19 pandemic on Asia-Europe relations. *Asia Europe Journal, 18*, 235–238.

Yu, Y., (2021). Resisting foreign hostility in China's English-language news media during the COVID-19 crisis. *Asian Studies Review*, 1–18.

CHAPTER 2

The Unsettling Rage and Fury Associated with Consuming Post-COVID-19 Space(s)

MAXIMILIANO E. KORSTANJE[1] and BABU GEORGE[2]

[1]*Department of Economics, University of Palermo, Larrea ST 1079, Buenos Aires, Argentina*

[2]*Alcorn State University, USA*

ABSTRACT

Over recent years, dark tourism has been a leading paradigm in tourism studies. Journalists, scholars, and the public have been captivated by the interplay between dark tourism and dark heritage. The recent commoditization of disaster discussed in Chapter 1 has invariably led to revising some open questions regarding what John Urry termed the tourist gaze. This book explores the strength and weaknesses of dark tourism literature as well as its potential applications in the new normal.

2.1 INTRODUCTION

At the turn of the century, the tourism industry was subject to countless disruptive events, which put its functionality in jeopardy (Laws, Prideaux & Chon, 2006; Aramberri, 2012; Ferreira & Visser, 2013; Robinson & Jarvie, 2008). Doubtless, the attacks on the World Trade Center, an event that shifted international relations forever, adjoined to many other risks

Post-Disaster and Post-Conflict Tourism (An Updated 2nd Edition), Maximiliano E. Korstanje, Hugues Seraphin, and Vanessa G. B. Gowreesunkar (Eds.)

as lethal virus outbreaks (like COVID-19), natural disasters, and the stock market crisis that whipped New York in 2008 resulted in the rise of anxieties and fears by the side of policymakers in tourism. As an industry of peace or at the best as intended to peacekeeping, tourism was essentially considered an activity that needs stability for prosperity (Somnez, 1998). This happens simply because tourists are afraid, the specialized literature said, to visit spaces where their integrity would be at stake, or in danger. This leads to the question of marketing sites of destruction for the consumption of tourists. While the consequences of this have been extensively studied by tourism researchers, some of the unintended effects have not yet been discussed: for instance, will the act of "making money out of disaster" reduce the perceived pain of the tragedy, especially when looking back at the disaster after an extended period? Will the opportunity to market disasters mean a tacit invitation to the conditions that can trigger disasters? If "yes" is the answer to such questions, there are more ethical issues with disaster marketing than what previously met the scholarly attention. Let's clarify to readers the recent COVID-19 pandemic as the main study case. Just after the first cases reported in Wuhan (China), the different governments disposed of a set of travel bans and restrictions to stop the virus proliferation. These measures included the closure of borders and airspace, as well as the imposition of strict lockdowns and social distancing, only to name a few. These restrictions not only affected seriously the tourism industry but also paved the way for the rise of a new tourist niche psychologically motivated to travel to the ground-zero, the market located in Wuhan where the first cases were sadly reported. This suggests two important things. On the one hand, stability and tourism safety are key factors towards a more secure destination, but on another, insecurity has attracted an ever-growing segment of tourists in recent years (Mansfeld & Pizam, 2006; Tarlow, 2014). Having said this, the doctrine of secure tourism sets the pace to a more tragic reality. The classic security programs not only failed to protect tourist destinations from terrorism and disasters, but also in some cases tourists were targeted as victims of irrational violence. As Korstanje (2017) puts it, it is almost impossible to imagine the end of tourism lest by the fact that terrorism is gradually eroding one of the tenets of Western civilization, hospitality. The decline of tourism rests on a manifest fear of alterity, which is systematically fostered by international terrorism. In the mid of this controversy, some voices have alerted on the emergence of a new trend, which commoditizes disasters as a form of popular entertainment (Stone & Sharpley, 2008; Dalton, 2014; Light, 2016). Dark Tourism, Thana Tourism, Slum Tourism, War Tourism, or even Prison tourism not only confronts the sand-and-beach products but also the Apollonian sense of beautiness, which historically characterized the classic tourism. Equally important, on the one hand, the emergence of dark tourism evinces the failure of risk-zero doctrine, as well as risk perception theory to prevent the negative effects of disasters (Korstanje, 2016). Secondly, tourism proved to be a resilient industry, or at least, a mechanism of resilience that helped society in the recovery processes (Korstanje & Ivanov, 2012; Crotti & Misrahi, 2015). While laypeople and audiences need to thematize death to domesticate the future and emerging anxieties, no less true is that these new forms of tourism offered alternative ways to channelize human solidarities towards a safer port (Sather Wagstaff, 2016). Recently, Hughes Séraphin and colleagues widely published on the importance of post-disaster marketing to include not only the anthropological discussion revolving around the nature of dark tourism but also the urgencies of integrating the academic knowledge to tourism management (Séraphin, Butcher, & Korstanje, 2017; Séraphin & Gowreesunkar, 2017).

In this introductory chapter, we discuss how tourism is a double-edged sword in post-disaster contexts: the support that tourism offers until suffering is commoditized for commercial advantage is a key focus. Ethically, though the gazing of others' pain may very well lead towards understanding and empathy when the situation is artificially fabricated and imposed two important risks emerge. We eventually forget the reasons for disasters, do not plan as well to prevent them, thereby increasing the probabilities the same event repeats. The conditions that made the disaster feasible are never corrected. To put this in other terms, Bianca Freire Medeiros reminds us how the ethical borders of slum tourism do not rest in the consumption of poverty or the miserable background where others live, but in the fact that "being poured in the global South" situates as the touchstone of this new style of gazing and the reasons behind why poverty is being replicated (Freire Medeiros, 2014). In this way, Tzanelli (2016) triggers a hot debate respecting the role of imperialism expansion of Western civilization. Through the articulation of a paternalist posture, the periphery was plundered of its resources while Europe accumulated further levels of wealth, which derived in an economic maturation. The consumption of *morbid esthetics* exhibits ironically the blame-others tactic where first-world tourists move freely to gaze at the misery, which has resulted from centuries of exploitation and center-periphery dependency. At the time, tourists interpret others' suffering as a new opportunity to understand their own lives; less is said of the neglected role of Europe in the pauperization of the world. We shall return to this theme shortly. The case of COVID-19 exhibits not only how fear works but also how it can be commoditized and packaged to be offered in the form of a dark spectacle to wider audiences.

2.2 THE SACREDNESS OF TOURISM

Emile Durkheim was one of the first pioneers in investigating the roots of religion as a myth-producer, as an ideological core from where rules and norms are orchestrated. Like his contemporaries, Durkheim believed that religion was a social construe, whose function was aimed at creating a source of solidarity among members of society. As a unified system of beliefs, religion bespeaks a supernatural order, which is the natural projection of social and economic forces. Durkheim not only considered religion as a fundamental institution but also as the source of society where collective consciousness is delivered. Tracing some ethnographies in Trobriand Islands and Australia, Durkheim found – though erroneously – that the religiosity within tribal societies was formed around the figure of the totem, which emanated all powers and authorities in the tribe. The advance of industrialism would wreak havoc in the social scaffolding accelerating some social pathologies, Durkheim assertively denounced as the necessary result for trust decline (Durkheim, 2008). The process of secularization was one of the main threats that concerned this French thinker. Although Durkheim centered his diagnosis on biased interpretations of old ethnographies and anthropological fieldworks, their contributions, as well as the figures he gave to sociology, are worldly valorized. Doubtless, one of the sociologists who devoted his life to the understanding of tourism and religiosity was Dean MacCannell. His assumptions are today widely cited by scholars worldwide, while his contributions shed light on the dichotomies of modern leisure. Dean MacCannell. In his classic work 'The Tourist,' MacCannell discusses to what extent modern societies should be conceptualized in opposition to tribal organizations. Here two assumptions are important. Firstly, in contemporary cultures, life is not centered on religion, or under the auspices of religious authority. The process of secularization accelerated the separation of state and religiosity, in a way that the gap is filled by the legal jurisprudence. In tribal cultures, instead, the source of any authority emanates from the totem, which is the religious figure that orders life. MacCannell takes its cue principally from the outcomes of Emile Durkheim and the illuminations of structuralism in social thought. Secondly, the social ties in secularized communities turn weaker than tribal units. The function of Totem in the tribal organization sets the pace for tourism for modern consumers. MacCannell innovatively proposes that in those capitalist societies, tourism plays a crucial role in keeping united the system of production. Whether Totems centralizes the authority of officialdom, through authenticity, museums provide visitors with a reason to live, giving a message to them. By the orchestration of cultural borders, the nation-state operates in a dimension of sacredness, where the quest for authenticity replicates the civic virtue of citizens. This point can be very well noted in the modes people travel worldwide. While our grandparents enjoyed their holidays always at the same destination, modern tourists need to gaze at something new every time. The notion of authenticity and novelty were vital forces of modern capitalism. Nonetheless, the dilemmas of ethics are coupled with the interests of empires, which often impose travels as the best of feasible realms. The possibility of being mobile, quite aside from culture or age, implies belonging to a privileged class that differentiates from the working-class. *In the 'ethics of sight-seeing,* 'MacCannell explores adamantly the connection of modernity with tourism. Confronted with the world of consumption, there is little place for ethics today. Equally important, scholars should not be so enthusiastic about the notion of heritage. It is time to reconsider the epistemology of heritage, as a meeting with others. At the same time, we watch others, these others are watching us. This reproduces a hermeneutic dialectics, where both sides negotiate their respective identities. But this engenders a serious risk when the code to interpret the otherness is previously fixed from outside. This prompts the creation of one-sided discourses which commoditize cultures, people, and their customs. The frenetic obsession for history, or by the instrumentalization of the past, alludes to the passage to new forms of living the sacredness. Once again, for MacCannell, if Totem centralized the power of sacred life in natives, tourism played the same role in a capitalist society, but unlike aborigines, the ubiquitous tourists visit remote lands in quest of what authenticity is. The tourist experience is based on the discovery of what is hidden (similarly to the gaze of medicine) (MacCannell, 1976, 2011). Since tourism needs suffering to offer the spectacle to spectatorship, no less true is that we have to change such a logic adopting a new ethic for tourism. The '*ethic of sightseeing*' reverses not only the climate of alienation in multi-ethnic encounters but allows tourists to put in the place of others (MacCannell, 2011). MacCannell does not raise the question to what extent dark tourism is the necessary proof to validate his theory, but he dangles that global modernism is undermining the autonomy of the alterity, according to a selfish drive of self-gratification and pleasure-maximization (Korstanje, 2016). Last but not least, Razaq Raj & Kevin Griffin (2015) offer an interesting discussion around the evolution of religious tourism to distinguish, what they dubbed as "the sacred" or "secular" journey. While the former prioritizes the curiosity to connect with others, the latter embraces religious beliefs and rules as the main guidelines for the trip. Equally important, they cite the advances of the anthropology of tourism, which toys with the belief tourism emulates a rite of passage. The discussion is given by the traces of Durkheim, who envisaged the intersection of religiosity and society. Raj & Griffin accept the Turnerian concept that the sacred world resonates with '*communitas.*' There would be a sacred-profane continuum in the religiosity of tourism which merits to be grasped.

2.3 TOURISM SPACE(S): CONFLICTS AND DISCONTENTMENT

Public opinion considers tourism as an industry that is oriented to keep peace and political stability. The multiplication factor provides local communities with fresh investments and employment but also with new opportunities to gain further independence from the central power. This is particularly true in those cases where hosting nations has a colonial dependency respecting first world countries or at best in those cases where tourism situates as the only source of production for a pauperized economy (de Kadt, 1979; Palmer, 1994). At a closer look, tourism can have positive and negative effects on local communities (Pizam, 1978; King, Pizam, & Milman, 1993). Among the positive effects, we have been the creation of jobs, which improve the conditions of life for locals, as well as the construction of infrastructures. In some contexts, tourism may help in protecting the environment (eco-tourism) or heritage (Ross, 1992;

Andereck et al., 2005). At thc same time, the industry brings with negative aftermaths, which include the saturation of the site, which leads to pollution and contamination, quality life decline, new emergent forms of alienation and the erosion of social ties (Archer, Cooper, & Ruhanen, 2005; Cunha, 2010). Certainly, some scholars have questioned the reasons why development failed to improve some third world economies. As de Kadt (1979) puts it, the colonial legacy, which means the historical dependency of colonial cultures from the central European matrix, resulted in serious social pathologies such as civil wars, corruption, and political instability that prevented the consolidation of host economies. In earlier studies, Korstanje alerted on the conceptual problems of the West to understand the "non-western other;" the theory of development, which is intended to preserve local cultures in the bias of extinction, are not pretty different from the colonial ethnocentric discourse that molded anthropology and ethnology. The quest for novelty and the obsession for discovery is not news. Either tends flourished in the 19th century, where European powers launched to domesticate other continents. The indexation of new economic resources for central economies was accompanied by the surge of an uncanny literacy genre fully dedicated to narrating the travelers' experiences as well as those obstacles they faced. This literature opened the doors for an ethnocentric view where Europe was enthusiastically enthralled as the peak of civilization, an ideal example to be implanted across the world. While Europeans felt the divine mandate to export their civilization, no less true was that the advance of European lifestyle would accelerate the disappearance of aboriginal cultures. Although from their outset, ethnology, and anthropology entered the field intending to salve natives, their outcomes, which provided accurate information, involuntarily benefited the colonial officers to colonize the alterity creating a vast far-flung empire. Not only development but also the contemporary theory of sustainability derives from old concerns which mask a subtle "paternalist stance" which merits to be discussed (Korstanje, 2012, 2017).

As the previous argument is given, Comaroff & Comaroff (2009) anticipated that colonialism forged a much deeper ideological discourse that not only introduced European rationality into the aboriginal minds but also applauded instrumentality as the supreme ideal to follow. Such a discourse persisted to date. The Comaroffs discuss the effects of heritage consumption in the periphery critically. Far from what popular opinion precludes, tourism sometimes engenders serious conflicts and discrepancies that usher natives into a civil war. While Aborigines are historically relegated by the nation-state, tourism revitalizes the economies of local communities. Paradoxically, though it acts as a mechanism towards poverty relief gaining further autonomy, the central administration often imposes heavy taxes over these thriving projects. In consequence, the reluctance of ethnic minorities to accept these policies opens the doors to extreme tension, which unless dully regulated culminates with civil wars or ethnic cleansings. This raises a more than an interesting question; i.e., why is the role of development and tourism in the configuration of a post-conflict culture? Is tourism part of the problem or the solution?

Though indirectly, Professor Phillip McMichael, in his book "*Development and Social Change,*" addresses the questions mentioned above. He toys with the idea that development continues the sentiment of the supremacy of old colonialism by other subtler means. The theory of globalization and development are inextricably intertwined. While the former signals as the "free market" as the best solution to all human evils the latter goes in the direction of protection. The protection of interests of global powers consists not only in securing the food production (in the South) to be exported to the North but also in the set of loans to keep "the market integration." The key factor of neo-liberalism is "governance," which means the coordination of NGOs by accessing information and material resources to fulfill the gaps left by "failed states." Today, corporate outsourcing is the crucial point Market is used to determine the contours of states. McMichael holds the thesis Europeans established a dual standard of morality once colonized the new world. On the one hand, they cultivated democracy and freedom as a mainstream cultural value, but at the same time, on another, they exerted violence systematically to discipline natives. Cruelty, submission, and violence were applied in the colonies, while in the core democracy prevailed as a valid system of government. This opens the doorstep to the process of "decolonization," where thousands of peripheral voices claimed to access the same rights "the democracy of their white lords" declared. McMichael explains that imperial powers alluded to the theory of "development" to maintain the dependency between the center and its periphery. The WWII end conjoined to Truman's administration led the United States to implement a wide range of credit systems to save the world from Communism. This program mushroomed to become in the development theory. As a mega-project, the theory of development was coined in 1940 and lasted until 1970. It not only created a food dependency but also accelerated the slum-dwellers and poverty in the peripheral countries. For remaking the old division of labor, Imperial Powers induced the "Third World" to accept international loans, which were used to industrialize their economies. At the time, under-developing nations adopted capital-intense methods in agriculture ruining the condition of small farmers, who migrated to urban areas, the US and Europe exported industrialized products. It was unfortunate the effects development left in Africa. The old boundaries of ethnicities the first colonial powers found were never honored once WWII was finalized. Many human groups were forced to live together within the fabricated limits of new nation-states. This resulted in a lot of ethnic cleansing, conflicts, and warfare that obscured the original ends of financial aid programs issued by the IMF or World Bank. Undoubtedly, the inconsistencies of the World Bank in administering the development-related programs not only were admitted but also woke up some nationalist reactions in the non-aligned countries. To restore the order, a new supermarket revolution surfaced: *globalization.*

Against this backdrop, McMichael writes that globalization expanded according to the lack of protective barriers in the Third World but prompted the decline of unionization in the industrial nations. The doctrine of free enterprise, which was presented as a superior ladder in the evolutionary process, fostered investors to fly to other continents and marketplaces, creating a divestiture process. Each state should adopt a specialized role in a much wider "world factory" where some provide the raw materials and others with elaborated products. These trends which characterized the 90s decade, created a new asymmetry between skilled (located in the first world) and under-skilled human resources (situated in the periphery). The recession produced by the oil embargo pressed First World to borrow a massive influx of money to the Third world, but now it will be carefully selected by two organisms, GATT and WTO. Both curtailed the protective measure of local economies by consolidating a new model which combined the reduced public capacity with the needs of governance. If Nationalism showed the importance of the nation-state to protect the citizen from the Market's arbitraries, now neo-liberalism focused on the inefficiency of public administration to regulate the economy.

> *"In short, the making of a free trade regime reconstructed food security as a market relation, privileging, and protecting corporate agriculture*

> *and placing small farmers at a comparative disadvantage. Food security would now be governed through the market by corporate, rather than social criteria" (p. 136).*

In this token, the World Trade Organization was relentless by charging or applying sanctions over those countries which affected the new system of import-export. Less interesting in freeing trade than in consolidating their hegemony, main powers prompted the discourse that Third World had not the right towards "self-sufficiency" anymore. Once again, globalization accelerated the accumulation of profits (beyond the boundaries and control of nation-states) but enlarged the levels of poverty as never before. Per some calculations, the UN declared that only 20% of the world population is situated within the 20% of the richest people, whereas the rest is facing serious economic problems. It is important not to lose sight that almost 1 billion is living in slums. As McMichael observes, the problem of poverty, far from being reverted, was aggravated, paving the way for internal riots, warfare, and conflicts. Last but not least, the colonial order has created a gap between having and have-nots, laying the foundations for serious material asymmetries in the under-developed nations. To this extent, some voices warn that the state of poverty, which reigns in many underdeveloped worlds in Northern Africa and the Middle East are fertile grounds for the cultivation of radicalized minds which can be later captivated by terrorism.

2.4 SPACE, SECURITY, AND PEACEKEEPING

Doubtless, tourism is an industry that fosters peace, encouraging integration between people and thus facilitating peaceful coexistence by respect and mutual tolerance. After all, travels not only undermine the action of racism and prejudice but also broaden the levels of ambiguity tolerance of the travelers (Jafari, 1989; Kim, Prideaux, & Prideaux, 2007). However, at a closer look, the opposite seems to be true; there is no consensus in pointing out that tourism improves the local quality of life, promoting internal stability which prevents the advance of violence (Ap & Var, 1990; Mansfeld, 1999). Yoel Mansfeld (1999) has widely studied the cycles of peace and wars in Israel, as well as the factors that affect tourism in Israel. Per se tourism is not an agent of peace, but once war has ended, it can help in the recovery process reconstructing cities that have been effaced. In addition, Pizam, Fleisher, & Mansfeld (2002) find that tourism, under some conditions, may change the previous prejudices, ideas, and stereotypes, changing the opinion respecting to the alterity. In this vein, Noel Salazar (2006) clarifies that peace is not only very hard to grasp but as an operating concept is almost impossible to measure. Tourism activates some resources to curb conflicts or give further insight into the protocols for conflict resolution. In any way, the idea of tourism as an instrument of peace should be reconsidered. As a force towards peace, tourism helps in improving the environmental quality, cultivating respect for others, as well as the overseas students' exchange. The correlation of tourism as a peacekeeper cannot be replicated in all cases. As Salazar notes:

> *"Empirical testing has not always supported the peace-through-tourism thesis and some scholars have argued that tourism seen as a force for peace is a "simplistic interpretation of the complexities of tourism and international relations." While often a co-relationship is found, with tourism as a beneficiary and/or consequence of peace, it is hard to prove the causal relation that tourism is a generator of peace. It is also difficult to make the case that tourism can prevent conflict. Before both World War I and World War II, for example, there was considerable private travel and tourism between the future combatant nations. However, travel and admiration of each other's cultures did not prevent war because the strength of personal sentiments was outweighed by political considerations. Of course, most tourism is dependent on peace and security. In a context of relative peace, logistical barriers to travel and psychological notions associated with fear for personal safety and antipathy from prospective hosts are removed and tourism is facilitated." (Salazar, 2006, p. 325).*

This cited paragraph ignites a hotly debate revolving around tourism in a widespread global order, where peace is very difficult to achieve. Raoul Bianchi and Markus Stephenson present a seminal book which is entitled "*Tourism and Citizenship.*" Their common-thread argument points out politics look to divide (polarize) cultures into two main blocs. For the sake of clarity, the Cold War pitted the US against the Soviet Union. Tourism never contributed to overcoming the barriers between these two worlds, but in enhancing nationalism internally. Like imperial travel in former centuries, tourism draws the borders of the civilized world and the barbarism. At the same time, tourism grew exponentially, becoming one of the most salient industries of the world, the same asymmetries which were enrooted in the capitalist ethos expanded. In this respect, far from bringing more peace, the process of globalization adjoined to tourism, engenders conflict. Tourism should be defined as something else than an economic industry, but it is a disciplinary instrument to domesticate others, strengthening the interclass inequalities. It is important not to lose sight of terrorism resulting from the resentment which is cemented by decades of subtle economic exploitation promoted by tourism. Tourists are killed and targeted as reminders of the political nature of tourism. As authors brilliantly observe, tourism incites rebellion since its political roots place guests above hosts. The host-guest encounter, which is based on subordination, infers the urgency of security. Nobody may travel abroad without the minimum conditions of protection, as Bianchi and Stephenson adhere, in the same way, that nobody feels appropriate dying as long as it's holidays. Security plays a vital role in configuring the borders of a postmodern world, where the logic of hyper-consumption is cataloged as a sign of supremacy while poverty and local crime is deterred as proof of cultural degeneracy. One of the ideological aspects of imperialism consists in making to believe the periphery there are strong reasons to take part in "privileged people."

Tourism would offer today – like travels in colonialism – a world without borders where free trade and political stability prevail. In so doing, tourism poses international citizenship not only commoditizing cultures and customs but also people and packaging them in the form of individual experiences. This creates "a double-oriented" mobility, where some are welcomed (tourists) while others are rejected (migrants and asylum-seekers). In this debate, the hope of a safer world still is a utopia simply because the current efforts of nations to securitize their homelands appear to be a clear invitation to terrorists. Equally important, the theory of risk perception, which was originally designed to prevent dangers earlier than they happen failed and its place was filled by a new theory, which focuses on dark consumption as a rapid and alternative way towards reconstruction. Based on the case of activist-artist Ai Weiwei, Tzanelli (2017) explains the artist agency, even the dark has some emancipatory potentialities. After years of struggle as activists, Weiwei denounced the human rights violation of the Chinese government as well as the lack of constitutional rights for dissidents. His artworks not only enact the darkness of politics but also entail a radical rejection of morbid consumption. China debates between the unwritten norm, which usually accepts the stranger-dissident in its terms and written rules which often evince a manifest hostility between documented dissidents and undocumented aliens. Far from being an absurd position or an insensitive artist who takes advantage of others' suffering, as Tzanelli clarifies, Ai replicates the points in discrepancy to offer an ethical view of hospitality. In this respect, the image of technology as an alienator force that disciplines citizens are singled out. Ai helps in forging a counterargument against the critiques on dark tourism. After all, dark consumption allows the production and free circulation of narratives, which escape often to human grasp, dealing with issues of "emotional reflexibility."

2.5 MORBID CONSUMPTION AND DARK TOURISM

Dark tourism has been widely addressed by different scholars, disciplines. Academicians of all stripes are captivated by dark tourism as an emergent field of research. It is particularly important to note that the interests for this issue have been directly proportional to an irreversible fragmentation of produced knowledge. The term '*dark tourism*' alludes to the visits to places characterized by mass death and mourning, but the same can be applied to other segments as Thana-Tourism, War-Tourism, Battle-Tourism, Slum-Tourism or even Prison Tourism. At the bottom, it is the same phenomenon though dotted with different shapes (Hooper & Lennon, 2016). One of the most authoritative voices in dark tourism, Phillip Stone argues convincingly that visitors intend to imagine their death, their finitude through the others' death. Through '*Thanaptosis,*' Stone adds, visitors, create empathy respecting to the others' pain, which helps to understand and learn from the trauma (Stone & Sharpley, 2008; Stone, 2012). As Cohen puts it, moral disasters explain society, which is rememorized through the consumption of dark tourism sites. In this process, people want to consume authentic experiences, looking for the soil where the victims fell, or in other cases, recreating the atmosphere of panic, before the event. Whatever the case may be, for Cohen, dark tourism serves as a pedagogic platform to explain to others the reasons for disasters (Cohen, 2011). Although from different perspectives, Cohen and Stone erroneously suggest that dark tourism is not new. In medieval times, peasants devoted considerable efforts and time traveling to be nearer to Saint's tombs. This is the empirical proof not only does dark tourism corresponds with a universal and anthropological drive, which means the need of domesticating death, but also it can be labeled as part of heritage, *dark heritage* (Stone, 2012; Cohen, 2011).

As a point of departure for many other studies, dark-tourism-as-heritage intends to show that psychological trauma helps the community not only to face the adversity but also give lessons (which are transmitted from generation to generation) to others who have never experienced such a situation. In this vein, dark consumption is part of the heritage that gives identity to locals (Biran, Poria, & Oren, 2011; Podoshen, 2013; Hartmann, 2014; Biran & Poria, 2012). Here two assumptions deserve to be done. On one hand, dark tourism seems to be part of a much broader obsession for death, which is enrooted in the capitalist system. By gazing at the suffering in other contexts, the public audiences reinforce their loyalties to nationhood, state, and the cultural values of their societies. At this point, dark tourism would serve as an ideological instrument of control. On another, no serious evidence validates Stone's insights medieval pilgrims visited dark sites for the same reasons as modern tourists. The motivations of holy pilgrims were historically associated to sacrifice or the preparation for the war, while tourists are only moving by curiosity or hedonist motives (Korstanje, 2016). The misleading confusion rests on the fact that supporters of heritage conducted open and closed-led questionnaires to visitors who manifested their needs of consuming heritage. This does not mean that dark tourism resulted from the medieval pilgrims. Second and most important, extrapolating the gathered information in interviews as a valid truth is a clear-cut error that puts the cart before the horse. This happens simply because sometimes interviewees lie (to protect their interests or to be complacent with the fieldworker), but in other cases, they even are incognizant of their emotional world (Korstanje, 2016). Joy Sather Wagstaff (2016) criticizes the role of heritage in tergiversating the real reciprocity trauma wakes up. Heritage is a distortion, a simplification of what trauma generates in human beings. The attacks on World Trade Center united not only all Americans but also the world. Americans received the solidarity and support of many allies, in the same way, distanced nations convey their sympathies. This reflects that death place humans in the same egalitarian conditions. With the passing of time the ground-zero, a theatralization of the event, was the opposite, allowing the tergiversation of 9/11 according to the demand of politics. For Wagstaff and other critical voices, heritage exhibits a cultural trend prone to commoditize the disaster, distorting its nature and opening the probabilities it can be repeated at a later day (Wagstaff, 2016; Korstanje, 2016; Tzanelli, 2016).

In her project, '*Mega-Events as Economies of Imagination,*' Tzanelli (2018) dangles that the spectacle capitalism proffers do not relate to the conditions of exploitation colonialism firstly and capitalism have been built, but to an externally-designed image which is imposed as the best of deserved worlds. This is possible because capitalism articulates two different mechanisms: the *artificial economy and the economy of imagination.* While the former refers to the normalcy of home and stability, the latter appeals to the internalization of universal values which interrogate the proper identity. As a result of this, the material background that facilitated the disaster is hidden, packaged, and sold as a spectacle that – undermining the critical thought – reinforces a nationalist posture. Since the economy of imagination draws the contours of ideology, the exploited classes are invited to take part temporarily in the wonderland only consumption offers. The state of pauperization, in many marginal countries, that left colonialism is not discussed nor do first-world tourists assume the ethical responsibilities by the actions of their state. Capitalism never corrects the glitches that imbued thousands of citizens into poverty. Rather, a morbid culture of consumption replicates the suffering as a commodity to be globally exchanged (Tzanelli, 2016). The question of whether social sciences were allies of colonial power remains in debate but what is true, capitalism abandoned the paradigm of the precautionary principle, which postulated the thesis that rationality suffices to prevent the emergence of risk, to transform the products of exploitation such as death, misery, slums, and poverty into embellished products. This moot point was recognized by Brazilian researcher B. Freire Medeiros in her ethnographies of slums in Rio. Slum tourism promotes a new form of intimacy between hosts and guests, but at the same time, a philosophical quandary. Since this segment needs from poverty to survive, there is no clear to what extent it helps locals to exit from slums and poverty (Freire Medeiros, 2014). Otherwise, slum tourism replicates poverty as the main commodity while the ruling elite keeps its status quo. To a major or minor degree, this exactly applies to morbid consumption. MacCannell is not wrong when he interposes the need of introducing ethics to tourism consumption (MacCannell, 2011).

2.6 THE GEOPOLITICAL TENSIONS IN A POST-COVID-19 WORLD

Doubtless, the COVID-19 pandemic not only captivated the attention of scholarships and audiences but also terrorized the world. No matter their GDP or levels of economic maturation, nations led invariably in an unparalleled crisis since their industries should be suddenly closed or stopped to their production. To the increasing unemployment generated by the strict lockdowns, one must add the collapse of sub-service sectors worldwide. In the mid of this mayhem, of course, the tourism industry was far from being an exception. As discussed in the introduction, the travel bans associated with the counter-measurements disposed to stop the virus dissemination placed the tourism industry on the brink of collapse. Having said this, the tourism industry was the main carrier and victim of the pandemic. The globalized world that historically characterized the West sets the pace to a new face of political disorganization and re-feudalization mainly marked by separatist movements, racism even hostility against Asian tourists. In some cases, the national borders were put into question and re-negotiated according to the new post-COVID-19 paradigm. Ultimately, tourists were demonized as potential carriers of a lethal virus. The tourist gaze, a concept originally coined by Urry to denote the fascination for the "Other" has been replaced by the Wicked Gaze, a new conception where all tourists become undesired guests! (Korstanje & George, 2021). Does this beg the question of whether specialists are witnessing now the end of tourism?

At a first glimpse, all appear to indicate that the tourism industry not only has been devastated by the COVID-19 but reborn from its ashes recovering rapidly to the earlier 2019 volume of passengers and travels. Nonetheless of this fact, the COVID-19 has altered not only the different documents or bans organized around the industry but also our current travel behavior. Tourists are now treated with some mistrust and fear to be potential silent carriers of COVID-19. It is safe to say that in this direction, as well as the hostility given against Asian Tourists should go the applied research in the years to come.

2.7 CONCLUSION

In summary, the post-disaster marketing theory aims to fix the conceptual problems of dark tourism theory while also offering a new methodological lens to examine the development of dark tourism promotional campaigns. Dark tourism is the result of a need felt in a society craving for the subaltern elements of existence – however, marketing plays a central role in giving voice to the solutions out there to meet the said need.

The landscape's esthetics are changing towards morbid forms. The successive failures of risk perception theory to protect the image of tourist destinations, accompanied with the European rational instrumentality to intervene in complex scenarios, contributed to the rise of dark tourism as a mechanism of resilience from where the community recovers in post-disaster contexts. In this introductory chapter, we have outlined the main limitations of dark tourism, calling attention to post-disaster marketing. As stated, one of the ethical limitations is exploiting the other's suffering convenes with the fact that unless solved the conditions that preceded the disaster may emerge. Post-disaster marketing does not correct the problem of war, violence, terrorism, or even climate change, but offers a valid toolkit to mitigate the negative effects of human negligence and indifference. Dark tourism and post-disaster marketing literature still captivate scholars given the ethical dilemma it poses. The idea of recycling post-conflict or disaster sites inscribes into an ever-changing climate of uncertainty where the traditional answers seem not to be enough. Here is where this chapter intends to contribute.

KEYWORDS

- **consumption**
- **COVID-19**
- **dark consumption**
- **dark tourism**
- **tourist gaze**

REFERENCES

Aaron, V. D., (n.d.). *Aesthetic Theory*. Retrieved from: http://www.aaronvandyke.net/summer_readings/Adorno-Hullot-Kentor.pdf (accessed on 20 January 2022).

Andereck, K. L., Valentine, K. M., Knopf, R. C., & Vogt, C. A., (2005). Residents' perceptions of community tourism impacts. *Annals of Tourism Research, 32*(4), 1056–1076.

Ap, J., & Var, T., (1990). Does tourism promote world peace? *Tourism Management, 11*(3), 267–273.

Aramberri, J., (2012). Is tourism vulnerable? *Tourism Recreation Research, 37*(3), 251–256.

Archer, B., Cooper, C., & Ruhanen, L., (2005). The positive and negative impacts of tourism. *Global Tourism, 3,* 79–102.

Bianchi, R., & Stephenson, M., (2014). *Tourism and Citizenship: Rights, Freedom and Responsibilities in the Global Order.* Abingdon: Routledge.

Biran, A., & Poria, Y., (2012). Reconceptualizing dark tourism. *Contemporary Tourist Experience: Concepts and Consequences*, 59–70.

Biran, A., Poria, Y., & Oren, G., (2011). Sought experiences at (dark) heritage sites. *Annals of Tourism Research, 38*(3), 820–841.

Cohen, E. H., (2011). Educational dark tourism at an in popular site: The holocaust museum in Jerusalem. *Annals of Tourism Research, 38*(1), 193–209.

Comaroff, J. L., & Comaroff, J., (2009). *Ethnicity*. Inc. Chicago, IL: University of Chicago Press.

Crotti, R., & Misrahi, T., (2015). The travel & tourism competitiveness index 2015: T&T as a resilient contribution to national development. *The Travel and Tourism Competitiveness Report, 2015*, 13.

Cunha, A. A., (2010). Negative effects of tourism in a Brazilian Atlantic Forest national park. *Journal for Nature Conservation, 18*(4), 291–295.

Dalton, D., (2014). *Dark Tourism and Crime* (Vol. 34). Abingdon: Routledge.

De Kadt, E., (1979). Social planning for tourism in the developing countries. *Annals of Tourism Research, 6*(1), 36–48.

Durkheim, E., (2008). *The Elementary Forms of the Religious Life*. New York, NY: Courier Corporation.

Freire-Medeiros, B., (2014). *Touring Poverty*. Abingdon: Routledge.

Hartmann, R., (2014). Dark tourism, Thanatourism, and dissonance in heritage tourism management: new directions in contemporary tourism research. *Journal of Heritage Tourism, 9*(2), 166–182.

Hooper, G., & Lennon, J. J., (2016). *Dark Tourism: Practice and Interpretation*. Abingdon: Routledge.

Jafari, J., (1989). Tourism and peace. *Annals of Tourism Research, 16*(3), 439–443.

Kim, S. S., Prideaux, B., & Prideaux, B., (2007). Using tourism to promote peace on the Korean peninsula. *Annals of Tourism Research, 34*(2), 291–309.

King, B., Pizam, A., & Milman, A., (1993). Social impacts of tourism: Host perceptions. *Annals of Tourism Research, 20*(4), 650–665.

Korstanje, M. E., & Ivanov, S. H., (2012). Tourism as a form of new psychological resilience: The inception of dark tourism. *Cultur: Revista de Cultura e Turismo, 6*(4), 56–71.

Korstanje, M. E., (2012). Reconsidering cultural tourism: An anthropologist's perspective. *Journal of Heritage Tourism, 7*(2), 179–184.

Korstanje, M. E., (2016). *The Rise of Thana Capitalism and Tourism.* Abingdon: Routledge.

Korstanje, M. E., (2017). *Terrorism, Tourism and the End of Hospitality in the West.* New York, NY: Springer.
Korstanje, M., & George, B., (2021). *Mobility and Globalization in the Aftermath of COVID-19.* Basingstoke, Palgrave Macmillan.
Laws, E., Prideaux, B., & Chon, K. S., (2006). *Crisis Management in Tourism.* Wallingford: Cabi.
Light, D., (2016). *The Dracula Dilemma: Tourism, Identity and the State in Romania.* Abingdon: Routledge.
MacCannell, D., (1976). *The Tourist: A New Theory of the Leisure Class.* Berkeley, CA: University of California Press.
MacCannell, D., (1984). Reconstructed ethnicity tourism and cultural identity in third world communities. *Annals of Tourism Research, 11*(3), 375–391.
MacCannell, D., (1992). *Empty Meeting Grounds: The Tourist Chapters.* London: Psychology Press.
MacCannell, D., (2002). The ego factor in tourism. *Journal of Consumer Research, 29*(1), 146–151.
MacCannell, D., (2011). *The Ethics of Sightseeing.* Berkeley, CA: University of California Press.
Mansfeld, Y., & Pizam, (2006). *Tourism, Security and Safety.* Abingdon, Routledge.
Mansfeld, Y., (1999). Cycles of war, terror, and peace: Determinants and management of crisis and recovery of the Israeli tourism industry. *Journal of Travel Research, 38*(1), 30–36.
McMichael, P., (2012). *Development and Social Change.* Thousand Oaks, SAGE.
Palmer, C. A., (1994). Tourism and colonialism: The experience of the Bahamas. *Annals of Tourism Research, 21*(4), 792–811.
Pizam, A., (1978). Tourism's impacts: The social costs to the destination community as perceived by its residents. *Journal of Travel Research, 16*(4), 8–12.
Pizam, A., Fleischer, A., & Mansfeld, Y., (2002). Tourism and social change: The case of Israeli ecotourists visiting Jordan. *Journal of Travel Research, 41*(2), 177–184.
Podoshen, J. S., (2013). Dark tourism motivations: Simulation, emotional contagion and topographic comparison. *Tourism Management, 35,* 263–271.
Raj, R., & Griffin, K. A., (2015). *Religious Tourism and Pilgrimage Management: An International Perspective.* Wallingford: CABI.
Robinson, L., & Jarvie, J. K., (2008). Post-disaster community tourism recovery: The tsunami and Arugam Bay, Sri Lanka. *Disasters, 32*(4), 631–645.
Ross, G. F., (1992). Resident perceptions of the impact of tourism on an Australian city. *Journal of Travel Research, 30*(3), 13–17.
Salazar, N. B., (2006). Building a 'culture of peace' through Tourism: Reflexive and analytical notes and queries. *Universitas Humanística, 1,* 319–336.
Sather-Wagstaff, J., (2016). *Heritage that Hurts: Tourists in the Memoryscapes of September 11.* Abingdon: Routledge.
Séraphin, H., & Gowreesunkar, V. G., (2017). Introduction: What marketing strategy for destinations with a negative image? *Worldwide Hospitality and Tourism Themes, 9*(5), 496–503.

Séraphin, H., Butcher, J., & Korstanje, M., (2017). Challenging the negative images of Haiti at a pre-visit stage using visual online learning materials. *Journal of Policy Research in Tourism, Leisure and Events, 9*(2), 169–181.

Sönmez, S. F., (1998). Tourism, terrorism, and political instability. *Annals of Tourism Research, 25*(2), 416–456.

Stone, P. R., (2012). Dark tourism and significant other death: Towards a model of mortality mediation. *Annals of Tourism Research, 39*(3), 1565–1587.

Stone, P., & Sharpley, R., (2008). Consuming dark tourism: A thanatological perspective. *Annals of Tourism Research, 35*(2), 574–595.

Tarlow, P., (2014). *Tourism Security: Strategies for Effectively Managing Travel Risk and Safety*. Oxford: Elsevier.

Tzanelli, R., (2016). *Thanatourism and Cinematic Representations of Risk: Screening the End of Tourism* (Vol. 176). Abingdon: Routledge.

Tzanelli, R., (2017). Schematizing hospitality: Ai WeiWei´s activist artwork as a form of dark travel. *Mobilities.* doi: 10.1080/17450101.2017.1411817.

Tzanelli, R., (2018). *Mega Events as Economies of the Imagination: Creating Atmospheres for Rio 2016 and Tokyo 2020.* Abingdon: Routledge.

Visser, G., & Ferreira, S., (2013). *Tourism and Crisis*. Abingdon: Routledge.

CHAPTER 3

Educating Tourists at the Pre-Visit Stage

HUGUES SÉRAPHIN,[1] KAREN CRIPPS,[2] and SIMON M. SMITH[3]

[1]*Department of Marketing, Event Management and Project Management, The University of Winchester, Winchester, Hampshire, SO22 4NR, UK*

[2]*Department of Responsible Management and Leadership, The University of Winchester Business School, Winchester, Hampshire, SO22 4NR, UK*

[3]*Oxford Brookes Business School, Oxford Brookes University, Oxford, OX3 0BP, UK*

ABSTRACT

It is safe to say that global capitalism has created the conditions for a more liberal market as well as free interchange among nations. This particularly pivoted in the rise and expansion of tourism, but at the same time, it made things harder for the destinations in view of a climate of excessive competition. This chapter brings some reflection on how educating tourists to visit destinations is mainly marked by an ongoing state of political violence and instability. The thesis held by the authors stands for tourism education avoids long-dormant ethnocentric discourses based on the fear of traveling to so-called risky destinations. This begs the question to what extent behind the alarmism of tourism safety and security lies deep western ethnocentrism.

Post-Disaster and Post-Conflict Tourism (An Updated 2nd Edition), Maximiliano E. Korstanje, Hugues Seraphin, and Vanessa G. B. Gowreesunkar (Eds.)

3.1 INTRODUCTION

Within this vast and complicated world of tourism, there are a great variety of destinations. These can grow and develop in different ways. Economic liberalization can play a leading role in such a process (Hoskinsson, Eden, Lau, & Wright, 2000). Concentrating on what we may consider as 'uniquely situated' destinations, some voices claim that low socio-economic development is conducive to political instability, intense conflict and the lack of a national identity, which have led very well to economic crises, disasters, and the outbreak of diseases (Gould, 2011; Ritchie, Dorrell, Miller, & Miller, 2004). Emerging countries are mostly postcolonial, post-conflict, or post-disaster (PCCD) destinations (Bayeh, 2015; Gould, 2011). These nations can be classified in two main groups. On one hand, developing countries in Asia, Latin America, Africa, the Middle East; and secondly, transition economies in the former Soviet Union and China (Hoskinsson et al., 2000). Séraphin, Ambaye, Gowreesunkar, & Bonnardel (2016b) hold that tourism works fine as a strategy for economic development, but, in some conditions, the tourism industry stimulates particular growth in some emerging destinations (Holden, 2013). Nonetheless, emerging destinations, which look to situate as tourist destinations are vulnerable to external factors such as instability, political riots and other risks (Ritchie, 2004). These risks accelerate the decline of the destination while in some other conditions, destinations can even disappear from the tourism map altogether (Seddighi, Nuttall, & Theocharus, 2001). With this, tourism is often esteemed as a complex activity (Korstanje & Tarlow, 2012). In view of this, some destinations have adopted strategies to mitigate their problematic status (Korstanje, 2009; Alvarez & Campo, 2014; Tarlow, 2014), such as targeting the diaspora as potential tourists (Séraphin & Paul, 2015). To set a clear example, Haiti experienced some changes its Destination Marketing Organization (DMO) logo in 2012, to adapt its change of image in its marketing strategy (Séraphin, Ambaye, Gowreesunkar, & Bonnardel, 2016b). Needless to say, the purpose of a logo is not to simply correct the problem of a country's negative image. A logo is also representative of a destination identity (Riel & Balmer, 1997) and, as such, activates the destination to communicate with potential visitors and *in fine* convince them to visit the destination (Séraphin, Ambaye, Gowreesunkar, & Bonnardel, 2016a). What is more important, a DMO logo synthetizes the history and nature of the destination, leading to brand identification and therefore clearer brand management (Aaker, 1997, 2012). In consequence this can also impact on purchase intention, positive emotions (Holbrook & Hirschman, 1982), trust, brand equity (Keller, 1993), brand attachment (Ball & Tasaki, 1992), brand commitment and subsequently improves the performance of a destination (Rose, Merchant, Orth, & Horstmann, 2016; Séraphin et al., 2016b; Park, Eisingerich, Pol, & Park, 2013).

Apart from a logo, there may be other marketing material that can capture the essence of the destination, which can lead to brand identification, impact on purchase intention, conveying positive emotions, trust, brand attachment and commitment. In this chapter, it is planned to test the ability of information and communication technologies (ICTs) to see if that can be achieved. More specifically, the objective of this research is to design a step-by-step method that may potentially help in deciding the content of an online platform designed to educate consumers at a pre-visit stage and potentially increase visitors to emerging markets. This method may also potentially help in deciding whether an online platform is the most suitable to represent and promote a destination. This is carried out by applying the DRA (Dale Robinson Anderson) model to an existing online platform (detailed in the Methodology section later on). This model is the combination of two existing models: The Dale & Robinson (2001) model and the Anderson (2001 – cited in Walters & Mair, 2012) model (Figure 3.1).

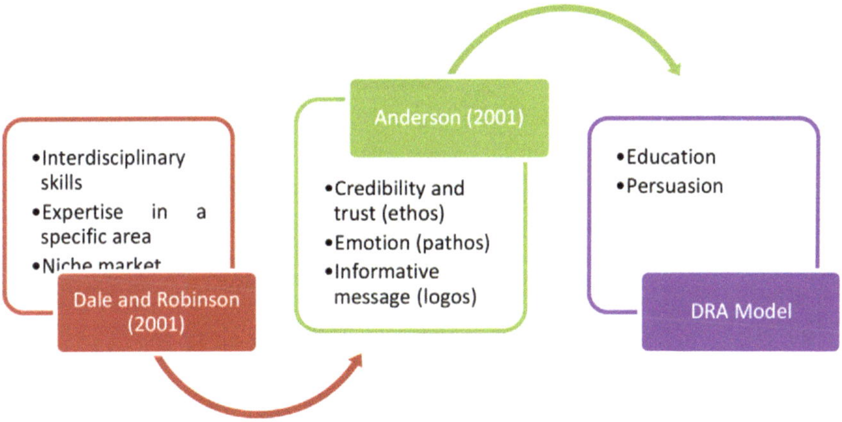

FIGURE 3.1 The Dale Robinson Anderson model (DRA model).

In this respect, Dale & Robinson (2001) forwarded a three-domain model of tourism education and an action plan for its implementation that would meet the needs of all stakeholders (learners, training providers, and employers). As for Anderson (2001 – cited in Walters & Mair, 2012), he put together a framework to test the effectiveness of a disaster recovery message drafted by a DMO. In this chapter, it is argued that a combination of the two models can provide the most suitable 'recipe' for a tool that can both educate people about an emerging destination that suffers from a negative image, and at the same time, encourage people to visit the destination. To some extent, this chapter is in line with Zhang (2000) who developed a model of quality management method on the basis of an extensive literature review.

Explicit and aggressive marketing strategies (e.g., planned communication activities, organization of international sports and cultural events, media strategies, exhibitions, changing the target audience, reducing hotel and flight prices) have proven to be ineffective for destinations suffering from a legacy of political instability and disaster (Séraphin et al., 2016a; Avraham, 2015; Alvarez & Campo, 2014; Seddighi et al., 2001). For this reason, this research has opted for a more subtle strategy, namely through 'education' using online materials.

With recent advances in ICTs, there is an interest among marketing researchers and managers to try to understand the challenges and opportunities for emerging markets (Sharma, 2011). This research suggests that education of tourists (Ballengee-Morris, 2002; Orams, 1996) at the pre-visit stage, using new technology as an opportunity for these markets, may contribute to improving a destinations' image. Existing research on tourists' education focuses on the visit itself, but generally neglects the pre-visit stage (Séraphin, Gowreesunkar, & Ambaye, 2016a). Also, the role of technology in tourism education, and their knowledge of the destination is currently little understood (Neuhofer & Buhalis, 2012). However, it has been shown that the internet provides many opportunities to disseminate knowledge (Liburd & Hjalager, 2010). Compared to more traditional forms of educational delivery, e-learning can improve student understanding of instructions, which in turn can enhance their motivation to learn and subsequently increase their emotional intelligence (Behnke, 2012). Behnke & Greenan (2011, cited in Behnke, 2012, p. 13) defined this as: "A measure of one's ability to use acquired knowledge, abstract thinking, and problem-solving to interpret and guide personal responses to significant internal and external situations." As for Computer-Assisted Instruction in hospitality management education, it can contribute to stimulating student participation and interest (Law, 2013).

This research can, therefore, be considered as a *continuation* of: (i) the research of marketing strategies for developing countries (Zarantonello, Schmitt, & Jedidi, 2014; Wei, Samiee, & Lee, 2014); (ii) the use of internet and social media as an advertising tool (Berthon, Pitt, Plangger, & Shapiro, 2012); and (iii) product innovation (Sok, O' Cass, & Miles, 2015) when it comes to the education of potential tourist at pre-visit stage (Séraphin, Butcher, & Korstanje, 2016c).

This research can also be considered as a *prolongation* of (iv) Alvarez and Campo's research (2014) on the impacts of political conflicts on a country's image and prospective tourists' intentions to visit. Indeed, it has been suggested that further research should be carried out to fully understand the image change process for countries that 'suffer from the chronically negative image and who are constantly exposed to political conflicts and other incidents' (Alvarez & Campo, 2014, p. 76).

This research can also be considered as a *complement* of (v) Avraham's (2015) research which argues that restoring the positive image of a destination relies on applying practical measures such as a marketing and sales strategy suitable to the context and situation of the destination.

Finally, this research could also be considered as a *continuation* of: (vi) Walters & Mair's (2012) research, who not only argued that there is limited literature in the area of post-disaster destinations, but who also emphasized the importance of further research in the field of post-disaster messages communicated via broadcast media, internet, social networking sites.

3.2 MARKETING STRATEGY OF DESTINATIONS WITH A NEGATIVE IMAGE

For many destinations, tourism is perceived as a panacea for economic and social development and hence, countries having a strong reliance on tourism invest considerably in visibility, marketability, and appeal of the destination. That said, not all destinations are vested with the same image. Because of this discrepancy, not all of them are performing at the same level. A study conducted by Séraphin, Gowreesunkar, & Blakeley (2016) on under-performing destinations showed that cognitive image formed on such destinations are derived from both internal and external sources; internal sources being from friends, peer and relatives and external sources being from media and public institutions. Likewise, another study by Gowreesunkar & Sotiriades (2017) indicated that image associated with underperforming destinations was termed as negative attributes in the tourism decision-making process and hence, significantly influenced the choice of destinations. According to the study, the image of a destination is an important attribute, as satisfaction and dissatisfaction may occur as a result of the customer's overall evaluation of the information associated with the destination' image. As a result, the type of information attached to a destination contributes to the image formation of a given destination. It is generally argued that marketing and more specifically branding plays an important role in overcoming some of the barriers related to the image. However, a study of Séraphin, Ambaye, Gowreesunkar, & Bonnardel (2016) probed that the process of branding of a destination is very hard to grasp in view of the ambidextrous approach which should be adopted. Destinations with negative images evolve in many ways and the marketing is very difficult to manage. To understand this, we need to mention that the specialized literature has not studied the functioning of image as well as its evolution in the threshold of time. As Gartner (2000) puts it, the destination image may be defined as 'the attitude, perception, beliefs, and ideas one holds about a particular geographic area formed by the cognitive image of a particular destination' (Gartner, 2000, p. 295). As for Cooper & Hall (2008, p. 223), the destination image seems to be a singled version of reality and it affects both an individual's perception of a destination and their choice of destination as the image of the destination is the only evidence that the visitor has of the destination before they visit. Both definitions highlight the importance of image in the decision making process to visit or not a destination. Even more important, no matter how the tourist acquired the image of the destination, but from his point of view, the perceived image is reality (Cooper & Hall, 2008, p. 224). Based on the above, and particularly on the fact that the media play a major role in the way a destination is perceived, it becomes quite obvious that destination with a negative image and destination with a positive image can't use the same marketing approach. Moreover, technology-savvy customers readily capture negative information which reaches through real-time, and they share to the online network. Once again, taking the example of Haiti, Séraphin, Butcher, & Korstanje (2016) explains that in this respect, Haiti does not have a 'high profile.' The shantytowns, the coups, etc., are generally the types of places and events broadcasted. The title of this theme, which is also the research question of this theme, makes it clear that there is a difference in the marketing strategy of destinations according to their image. The nature of the tourism industry makes it even more difficult for a destination with a negative image 'to rise from the ashes.' Indeed, the tourism sector is Those factors can cause destinations to decline and sometimes even totally disappear from the tourism map (Avraham, 2015; Séraphin, Butcher, & Korstanje, 2016; Alvarez & Campo, 2014). As elaborated above, technology-savvy customers of the 21st century are now increasingly resorting to online sources to acquire and share information on destinations. Therefore, identifying a marketing strategy for destinations with a negative image is not the main issue per se, rather it is the type of information created on the destination and the medium to be utilized to launch or re-launch the destination that needs to be identified and worked upon. The crafting of appropriate marketing strategies for destinations with poor image needs to take into account various factors, in particular, their online visibility and types of information shared.

Morgan, Pritchard, & Pride (2011) discussed a range of destinations (Singapore, Abu Dhabi, Macau, Milan, Australia, Namibia, San Antonio, Slovenia, London, New York, Barcelona, Northern Ireland and Jerusalem) with different histories, reputations, resources, markets, and levels of tourism development. In terms of branding strategies adopted to be competitive, three groups could be identified: (i) the first group that includes Singapore, Abu Dhabi, Macau, Milan, New York, London, and Barcelona adopted a strategy based on the organization of events and the development of landmark projects to gain the reputation of world-leading cities. (ii) As for Australia, it focuses its effort of film's theme to build up its image. And (iii) as for the third group (Namibia, San Antonio, Slovenia, Northern Ireland, and Jerusalem) they adopted a marketing strategy with a strong focus on authenticity, culture, and a sense of place. For all three groups, no matter their history, reputation, resource, market, and levels of tourism development opted for solution focus on the destination. They are traditional solutions that have so far proven ineffective (Alvarez & Campo, 2011, 2014; Avraham, 2015; Vitic & Ringer, 2008; Walters & Mair, 2012). In this theme, we are going to explore alternative solutions. Some will still focus on the destinations and others will focus on the tourist. The latter is of interest as a limited research focus on the tourists.

As the image of a destination is based upon induced agents, organic agents and autonomous agents (as explained previously), Séraphin, Butcher, & Korstanje (2016) argued the importance of educating tourists a pre-visit stage using virtual online learning material (VOLM) as educated potential tourists are less sensitive to adverse publicity as they develop a special bond with the destination. In this theme, we are going a step further by suggesting an ambidextrous approach by also suggesting education of the tourists during the visit as suggested by Orams (1996).

3.3 IMPROVING THE IMAGE OF POSTCOLONIAL, POST-CONFLICT, AND POST-DISASTER

The experience of the Arab Spring shows two important aspects. As Avraham (2015) observed, the fact is that tourist destinations in the Middle East deploy efforts in locking the action of the media, trying to transform the negative image into a positive one. The post-conflict strategies should be taken as a resource the promotional video-information impact, as well as the articulation of effective recovery messages (Walters & Mair, 2012; Alvarez & Campo, 2011). While other studies suggest that the involvement of the locals is a crucial factor that explains the success of the emerging destination (Ballengue-Morris, 2002; Orams, 1996).

All the examples and testimonies are centered on the education of tourists while visiting the destination. This chapter, rather, attempts to make a fresh approach focusing on the pre-visitation stage. The pre-visitation stage is an important phase of information processing and is shaped by personal knowledge and other third-party information (Hubner & Gossling, 2012; Séraphin et al., 2016c). The literature suggests that the image starts from the ambiguous premise that it is defined by the subjective impression of each person (O'Sullivan, 1996, cited in Seddighi et al., 2001). Such an impression appears significantly formed by the affective disposition instead of the cognitive structure (Alvarez & Campo, 2014). This process reinforces while confronts with the established media discourse while re-educate to the tourist (Séraphin, 2013; Reisinger & Steiner, 2006; Benhke, 2012). While tourism education is generally only seen from the students,' academics,' practitioners' and employers' angle (Dale & Robinson, 2001), this chapter considers it as a pre-visit marketing tool. This therefore explicitly aims to counter false or exaggerated negative notions relating to colonialism, disasters, and conflicts of the past (both distant and recent). When people have a better knowledge of the true nature of a destination (history, tradition, context, etc.), they are more likely to visit it because they will be less open to distortions, exaggerations, and unbalanced accounts (Gossling, Scott, Hall, Ceron, & Dubois, 2012).

Consumer behavior can be defined as: "behavior that consumers demonstrate whilst searching for, buying, using, evaluating, and disposing of services and products that are expected to satisfy their needs" (Schiffman, Kanuk, & Wisenblit, 2010). More specifically, and based on a tourism context, Kirillova et al. (2014) explains that elements that tourists consider as esthetically pleasing or beautiful (environment, product or services, tourism experience), can contribute to their decision to visit a destination, to return to the same destination and, in the long term, to show loyalty to that destination. The esthetic case for destinations with a negative image rarely reaches potential tourists due to media focus on risks arising from disasters, conflict, and the nation's postcolonial status (Séraphin & Nolan, 2014a).

Importantly for this study, though, Carlson's natural environment model (1979) and the arousal model developed by Caroll (1995) suggest that knowledge and objectivity about the environment can influence appreciation. Their analysis implies that objective knowledge and exposure to more balanced narratives, influence the way potential tourists approach destination choice not just on a cognitive/rational basis, but also on an affective one. Hence the education of tourists can be considered as a way to facilitate a more favorable affective disposition towards destinations with the negative image. Also, pertinent here is Connell's view (cited in Stepchenkova, 2013) that non-promotional communications can be more powerful than more explicitly promotional ones. For specific destinations, specific measures must be taken (Gay, 2012; Huang, Tseng, & Petrick, 2008, cited in Walters & Mair, 2012). Appealing to education and a desire for knowledge, rather than romantic imagery and persuasive prose, could be more effective in undermining the perception of PCCD destinations through a dark lens.

3.4 CONSCIOUSNESS, POWERS, AND DRIVERS FOR A SUSTAINABLE FORM OF TOURISM

Sustainability is the most pressing issue faced by the tourism industry and more generally speaking, it is the most important issue of this century (Higham et al., 2021), as a result it is shaping every aspect of consumers decision making

(Han, 2021). Being pro-environmental is a growing trend, even a living style for many consumers (Han, 2021). However, what is really needed for change to happen is a radical innovation that is going to disrupt current practices, as the incremental innovations implemented so far have failed to bring sustainable changes in the industry. A rethinking of what responsible tourism is and should be achieved need a total rethinking (Burrai et al., 2019). This radical change is likely to happen with the emergence of tourism sustainability activists (Séraphin, 2022; Tranter, 2010), and an education of both locals and visitors (Wang, Zhang, Xiao, Sun, Xia, & Shi, 2020). Even if that it has been noticed that when on holidays, individuals retain their pro-environmental behavior (Wang et al., 2020), sometimes in happens that this attitude change (Juvan & Dolnicar, 2014). Additionally, it is worth mentioning that despite the fact the education of tourists (and locals) at pre-visit and during the visit stage, plays a significant role in their behavior (Han, 2021), their level of pro-environment behavior will vary according to culture, age, gender, social background, education, income, health where they grew up, their personality, etc. (Abrahamse, 2019; Balundé & Kim, 2012; Dolnicar, Crouch, & Long, 2008; He & Filimonau, 2020; Perlaviciute & Steg, 2019; Séraphin, 2022). To develop this pro-environment consciousness and behavior, strategies which can be classified as coercive (use of power to force others to do something), legitimate (prescription of action plans), induced (financial resources), and competent (advice to others) are put in place by private and public sector organizations involved in the industry (Saito & Ruhanen, 2017). Having said that, trust is needed amongst all stakeholders for these strategies to be effective (Kikuchi-Uehara et al., 2016; Nunkoo, 2017). Another important element for the development of pro-environment behavior with customers is the development of an intrinsic, altruistic, and biospheric motivation, as opposed to an extrinsic, self-interested, and egoistic motivation (Evans et al., 2012; Sharma & Gupta, 2020; Wu, Font, & Liu, 2020). This intrinsic, altruistic, and biospheric motivation often comes with place attachment (Hong & Kim, 2019), hence why it is important for DMO to also adopt affective marketing strategies (Choi & Kim, 2021).

3.5 TOUR OPERATORS – SHAPING TOURISTS' AWARENESS AND ACTION

Tour operators play a powerful role in shaping tourist behaviors both pre and during visit, since they connect destination suppliers with consumers through bundling package components such as accommodation and transport. Online retail distribution has reduced the power of tour operators in controlling tourism flows that shape destination development and was widely criticized in literature at the turn of the century (see for example Medina-Munoz et al., 2003; Papatheodorou, 2006; Buhalis, 2000; Bastakis et al., 2004). Paradoxically however, it can be suggested that COVID-19 provides tour operators an advantage in providing assurance to consumers of health and safety procedures and can play a key role in helping to re-build tourism. Consumers can be faced with complex risk rating information that changes at short notice (see for example the advice issued by Public Health Scotland, 2022), and tour operators can assist consumers in providing assurances of the safety of travel arrangements. The Association of British Travel Agents' (ABTA, 2022) campaign #ReadySteadyTravel for example, provides information for consumers on what to consider when booking, and what to expect at the airport, during the journey and in destination. This merits further research into how consumer decision making post-COVID-19 has been influenced in terms of a desire to book through a tour operator, rather than independently based upon the information and support that tour operators can provide.

Tour operators also play an influential role in shaping tourists' pro-environmental behaviors through pre and during visit communication. Reflecting a wider paradigm shift towards more responsible and less exploitative forms of capitalism, even the largest online retailers can be seen to include information on 'sustainable' tourism. This is a significant change to the turn of the century when research identified that operators did not always take a long-term view to destination development and claimed to lack control over impacts in destination (Carey et al., 1997; Curtin & Busby, 1999; Forsyth, 1995; Klemm & Parkinson, 2001; Miller, 2001; Tapper, 2001). It is encouraging that larger mass-market players now communicate sustainable information alongside specialist businesses that are frequently founded on sustainable principles and have long communicated desirable behaviors to consumers. 'Travalyst' (2021, p. 14) for example which represents large global players such as Booking.com, Google, Skyscanner, Trip.com Group, TripAdvisor, and Visa, notes "Growing traveler awareness of sustainability. Travelers are already increasingly demanding more sustainable options, with more than two-thirds (69%) expecting the travel industry to offer more options that deliver on this desire."

'Tourism sustainability activists' are classified by Séraphin (2022) as being passive or active, depending on contextual (such as age, gender, education, and background) or person determinants (personality traits). Tour operators' communication of sustainability information to consumers (defined by Font & McCabe (2017) as 'market driven approaches' to sustainability), alongside sustainable product design ('product driven approaches') can be considered a key contextual influence on active sustainability activism. Effective consumer information to drive awareness and action is fundamental to tour operators sustainable supply chain management (SSCM) processes (Richards & Font, 2019; Schwartz, 2008). As the world transitions out of COVID-19 restrictive travel bans, the importance of consumer education is arguably more important than ever. Gossling et al. (2021) write of the risk of returning to 'business as usual,' perhaps following even more aggressive growth to compensate for losses to the tourism industry. It is evident that in a complex and challenging business environment, all communication 'touch points can shape pro-environmental behaviors and resilient business models. The shift set out in this section from the turn of the century, which was characterized by a denial of responsibility towards engagement, needs to gather pace through consistent focus on communication and knowledge sharing.

3.6 CONCLUSIONS

When educating tourists at the pre-visit stage, the chapter has shown there are a number of key factors to consider within marketing strategies. In destinations that have incredibly unique histories and heritages, e.g., post-colonial, post-conflict, and post-disaster, the emphasis on pre-visit education is potentially paramount to a tourist's decision-making process. Thus, marketing and online artifacts need to reflect this educational journey to obtain more buy-in from potential tourists and increase the potential for tourism within these destinations. Perhaps the crucial message to convey here is that it is possible to have successful tourism within these unique destinations – the evidence is there to support this.

The world has changed significantly since 2020 and the years that followed adapting to COVID-19. Tourism destinations have been badly affected through this time. Thus, the importance of finding recovery that turns into success is essential to so many destinations. On top of this, we have a society now, more than ever, focused on critical issues surrounding sustainability and sustainable tourism. This means that the tourism industry has needed great agility to cope with all these external demands and challenges. The destinations that can best manage the complexities are likely to be the ones achieving long-term growth and survival. There is a lot to consider and there are likely to be winners and losers within this process. Yet, we would argue that something effectively implemented like pre-visit tourist education can help to build successful marketing strategies.

KEYWORDS

- **adaptability**
- **dangers**
- **education**
- **political instability**
- **risk perception**
- **risky destinations**

REFERENCES

Aaker, D. A., (1997). Dimensions of brand personality. *Journal of Marketing Research, 34*(3), 347–356.

Aaker, D. A., (2012). *Building Strong Brands*. New York, NY: Free Press.

Abrahamse, E., (2019). Behavior change intervention. *Encouraging Pro-Environment Behavior*. doi: 10.1016/B978-0-12-811359-2.00003-2.

Abrahamse, E., (2019). *Understanding the Drivers of Human Behavior, Encouraging Pro-Environment Behavior.* doi: 10.1016/B978-0-12-811359-2.00002-0.

Alvarez, M. D., & Campo, S., (2011). Controllable versus uncontrollable information sources: Effects on the image of Turkey. *International Journal of Tourism Research, 13*, 310–323.

Alvarez, M. D., & Campo, S., (2014). The influence of political conflicts on country image and intention to visit: A study of Israel's image. *Tourism Management, 40*, 70–78.

Association of British Travel Agents, (2022). *Travelling During Corona Virus*. Available from: https://www.abta.com/tips-and-advice/travelling-during-coronavirus (accessed on 20 January 2022).

Avraham, E., (2015). Destination image repair during crisis: Attracting tourism during the Arab spring. *Tourism Management, 47*, 224–232.

Ball, A. D., & Tasaki, L. H., (1992). The role and measurement of attachment in consumer behavior. *Journal of Consumer Psychology, 1*(2), 155–172.

Ballengee-Morris, C., (2002). Cultures for sale: Perspectives on colonialism and self-determination and the relationship to authenticity and tourism. *Studies in Art Education, 43*(3), 232–245.

Balundė, A., Perlaviciute, P., & Steg, L., (2019). The relationship between people's environment considerations and pro-environmental behavior in Lithuania. *Frontiers in Psychology*. doi: 10.3389/fpsyg.2019.02319.

Bastakis, C., Buhalis, D., & Butler, R., (2004). The perception of small and medium sized tourism accommodation providers on the impacts of the tour operators' power in Eastern Mediterranean. *Tourism Management, 25*, 151–170.

Bayeh, E., (2015). The legacy of colonialism in the contemporary Africa: A cause for intrastate conflicts. *International Journal of Innovative and Applied Research, 3*(2), 2329.

Behnke, C., (2012). Examining the relationship between emotional intelligence and hospitality student attitudes toward e-learning. *Journal of Hospitality & Tourism Management, 24*(2), 12–20.

Berthon, P. R., Pitt, L., Plangger, K., & Shapiro, D., (2012). Marketing meets web 2.0 social media, and creative consumers: Implication for international marketing strategy. *Business Horizons, 55*(3), 261–271.

Buhalis, D., & Law, R., (2008). Progress in information technology and tourism management: 20 years on and 10 years after the internet—The state of e-Tourism research. *Tourism Management, 29*(4), 609–623.

Buhalis, D., (1998). Strategic use of information technologies in the tourism industry. *Tourism Management, 15*(5), 409–421.

Buhalis, D., (2000). Relationships in the distribution channel of tourism: Conflicts between hoteliers and tour operators in the Mediterranean region. *International Journal of Hospitality and Tourism Administration, 1*(1), 113–139.

Burrai, E., Buda, D. M., & Stanford, D., (2019). Rethinking the ideology of responsible tourism. *Journal of Sustainable Tourism, 27*(7), 992–1007.

Carey, S., Gountas, Y., & Gilbert, D., (1997). Tour operators and destination sustainability. *Tourism Management, 18*(7), 425–431.

Carlson, A., (1979). Appreciation and the natural environment. *The Journal of Aesthetics and Art Criticism, 37*(3), 267–275.

Choi, S., & Kim, I., (2021). Sustainability of nature walking trails: Predicting walking tourists' engagement in pro-environmental behaviors. *Asia Pacific Journal of Tourism Research, 26*(7), 748–767.

Curtin, S., & Busby, G., (1999). Sustainable destination development: The tour operator perspective. *International Journal of Tourism Research, 1*, 135–147.

Dale, C., & Robinson, N., (2001). The theming of tourism education: A three-domain approach. *International Journal of Contemporary Hospitality Management, 13*(1), 30–34.

Dixit, S. K., (2017). *Handbook of Consumer Behavior in Hospitality and Tourism*. Abingdon: Routledge.

Dolnicar, S., Crouch, G. I., & Long, P., (2008). Environment-friendly tourists: What do we really know about them? *Journal of Sustainable Tourism, 16*(2), 197–210.

Font, X., & McCabe, S., (2017). Sustainability and marketing in tourism: Its contexts, paradoxes, approaches, challenges and potential. *Journal of Sustainable Tourism, 25*(7), 869–883.

Forsyth, T., (1995). Business attitudes to sustainable tourism: Self-regulation in the UK outgoing tourism industry. *Journal of Sustainable Tourism, 3*(6), 210–230.

Gartner, W. C., (1993). Image formation process. *Journal of Travel and Tourism Marketing, 2*(2, 3), 191–215.

Gay, J. C., (2012). Why is tourism doing poorly in overseas France? *Annals of Tourism Research, 39*(3), 1634–1652.

Gössling, S., Scott, D., & Hall, C. M., (2021). Pandemics, tourism and global change: A rapid assessment of COVID-19. *Journal of Sustainable Tourism, 29*, 1–20.

Gossling, S., Scott, D., Hall, M. C., Ceron, J. P., & Dubois, G., (2012). Consumer behavior and demand response of tourists to climate change. *Annals of Tourism Research, 39*(1), 36–58.

Gould, M., (2011). *Branding a Post-Conflict Destination: Northern Island.* In Morgan, N.

Hall, C. M., (2002). Travel safety, terrorism and the media: The significance of the issue-attention cycle. *Current Issues in Tourism, 5*(5), 458–466.

Han, H., (2021). Consumer behavior and environmental sustainability in tourism and hospitality: A review of theories, concepts, and latest research. *Journal of Sustainable Tourism, 29*(7), 1021–1042. doi: 10.1080/09669582.2021.1903019.

He, L., & Filimonau, V., (2020). The effect of national culture on pro-environmental behavioral intentions of tourists in the UK and China. *Tourism Management Perspective*. doi: 10.1016/j.tmp.2020.100716.

Higate, P., & Henry, M., (2009). Insecur*e Spaces, Peacekeeping, Power and Power and Performance in Haiti, Kosovo and Liberia.* London: Zed Books.

Higham, J., Font, X., & Wu, J., (2022). Code red for sustainable tourism. *Journal of Sustainable Tourism*. doi: 10.1080/09669582.2022.2008128.

Holbrook, M. B., & Hirschman, E. C., (1982). The experiential aspects of consumption: Consumer fantasies, feelings, and fun. *Journal of Consumer Research, 9*(2), 132–140.

Holden, A., (2013). *Tourism, Poverty and Development.* Abingdon: Routledge.

Hong, J. J., & Kim, N. J. K., (2019). An investigation of the relationship between place attachment (PA) and pro-environmental behavioral intentions (PEBI) and its implications towards over-tourism. *2019 TTRA International Conference*.

Hoskinsson, R. E., Eden, L., Lau, C. M., & Wright, M., (2000). Strategy in emerging economies, *Academy of Management Journal, 43*(3), 249–267.

Hubner, A., & Gossling, S., (2012). Tourist perceptions of extreme weather events in Martinique. *Journal of Destination Marketing & Management, 1*(2012), 47–55.

Ioannides, D., & Daughtrey, E. P., (2006). Competition in the travel distribution system. In: Papatheodorou, A., (ed.), *Corporate Rivalry and Market Power: Competition Issues in the Tourism Industry* (pp. 124–142). London: I.B. Tauris and Co. Ltd.

Juvan, E., & Dolnicar, S., (2014). The attitude-behavior gap in sustainable tourism. *Annals of Tourism Research, 48*, 76–95.

Keller, K. L., (1993). Conceptualizing, measuring, and managing customer-based brand equity. *Journal of Marketing, 57*(1), 1–22.

Kikuchi-Uehara, E., Nakatani, J., & Hirao, M., (2016). Analysis of factors influencing consumers' pro-environmental behavior based on life cycle thinking. Part II: Trust model of environmental information. *Journal of Cleaner Production*. doi: 10.1016/j.jclepro.2016.03.011.

Kirillova, K., Fu, X., Lehto, X., & Cai, L., (2014). What makes a destination beautiful? Dimensions of tourist aesthetic judgement. *Tourism Management, 42*(2014), 282–293.

Klemm, M., & Parkinson, L., (2001). UK tour operator strategies: Causes and consequences. *International Journal of Tourism Research, 3*, 367–375.

Korstanje, M. E., & Tarlow, P., (2012). Being lost Tourism, risk and vulnerability in the post-'9/11'entertainment industry. *Journal of Tourism and Cultural Change, 10*(1), 22–33.

Korstanje, M., (2009). Re-visiting risk perception theory in the context of travel. *E-Review of Tourism Research, 7*(4), 68–81.

Law, R., (2013). Computer-Assisted instruction in hospitality education: An intelligent tutoring systems approach. *Journal of Hospitality & Tourism Education, 11*(1), 50–56.

Liburd, J., & Hjalager, A. M., (2010). Changing approaches towards open education, innovation and research in tourism. *Journal of Hospitality and Tourism Management, 17*, 12–20.

Medina-Munoz, R. D., Medina-Munoz, D. R., & Garcia-Falcon, J. M., (2003). Understanding European tour operators' control on accommodation companies: An empirical evidence. *Tourism Management, 24*, 145–147.

Miller, G., (2003). Consumerism in sustainable tourism: A survey of UK consumers. *Journal of Sustainable Tourism, 11*(1), 17–39.

NACETT, [Online], Available from: http://www.acronymfinder.com/National-Advisory-Council-for-Education-and-Training-Targets-(UK)-(NACETT).html (accessed on 20 January 2022).

Nunkoo, R., (2017). Governance and sustainable tourism: What are the role of trust, power and social capital? *Journal of Destination Marketing Management*. doi: 10.1016/j.jdmm.2017.10.003.

Orams, M. B., (1996). A conceptual model of tourist-wildlife interaction: The case for education as a management strategy. *Australian Geographer, 27*(1), 39–51.

Park, C. W., Eisingerich, A. B., Pol, G., & Park, J. W., (2013). The role of brand logos in firm performance. *Journal of Business Research, 66*, 180–187.

Public Health Scotland, (2022). *Fit for Travel*. Available from: https://www.fitfortravel.nhs.uk/news/newsdetail.aspx?id=24163 (accessed on 20 January 2022).

Reisinger, Y., & Steiner, C., (2006). Reconceptualizing interpretation: The role of tour guides in authentic tourism. *Current Issues in Tourism, 9*(6), 481–497.

Richards, P., & Font, X., (2019). Sustainability in the tour operator – ground agent supply chain. *Journal of Sustainable Tourism, 27*(3), 277–291.

Riel, C. B. M., & Balmer, J. M. T., (1997). Corporate identity: The concept, its measurement and management. *European Journal of Marketing, 31*(3, 5), 340–355.

Ries, A., & Trout, J., (1982). *Positioning, the Battle for Your Mind*. McGraw-Hill Professional.

Ritchie, B. W., Dorrell, H., Miller, D., & Miller, G. A., (2004). Crisis communication and recovery for the tourism industry. *Journal of Travel and Tourism Marketing, 15*(2, 3), 199–2016.

Rose, G. M., Merchant, A., Orth, U. R., & Horstmann, F., (2016). Emphasizing brand heritage: Does it work? And how? *Journal of Business Research, 69*, 936–943.

Saito, H., & Ruhanen, L., (2017). Power in tourism stakeholder collaborations: Power types and power holders. *Journal of Hospitality and Tourism Management, 31*, 189–196.

Schiffman, L. G., Kanuk, L. L., & Wisenblit, J., (2010). *Consumer Behavior* (10th edn.). London: Pearson Prentice Hall.

Schwartz, K., Tapper, R., & Font, X., (2008). A sustainable supply chain management framework for tour operators. *Journal of Sustainable Tourism, 16*(3), 298–314.

Seddighi, H. R., Nuttall, M. W., & Theocharous, A. L., (2001). Does cultural background of tourists influence the destination choice? An empirical study with special reference to political instability. *Tourism Management, 22*(2001), 181–191.

Séraphin, H., & Nolan, E., (2014). Voodoo in Haiti: A religious ceremony at the service of the 'houngan' called 'tourism.' In: Frost, W., & Laing, J., (eds.), *Rituals and Traditional Events in the Modern World.* New-York: Routledge.

Séraphin, H., & Paul, B., (2015). *The Diaspora: A Lever for the Development of Tourism in Haiti. Worlds of Tourism* [Online] http://tourisme.revues.org/990 (accessed on 20 January 2022).

Séraphin, H., (2013). *The Contribution of Tour Guides to Destination Understanding and Image. The Case of Haiti via an Analysis of: 'Bonjour Blanc, a Journey Through Haiti.'* International Research Forum on Guided Tours, Breda University of Applied Sciences (Netherland).

Séraphin, H., (2014). *Le Tourisme: L'ouverture Pour le Peuple de Toussaint?* Paris: Publibook.

Séraphin, H., (2022). Understanding the traits of tourism sustainability activists through a life course framework. *Journal of Policy Research in Tourism, Leisure and Events.* Ahead of Print 1–19. https://doi.org/10.1080/19407963.2022.2029873.

Séraphin, H., Ambaye, M., Gowreesunkar, V., & Bonnardel, V., (2016). A marketing research tool for destination management organizations' logo design. *Journal of Business Research.* doi: 10.1016/j.jbusres.2016.04.074.

Séraphin, H., Butcher, J., & Konstanje, M., (2016). Challenging the negative images of Haiti at a pre-visit stage using visual online learning materials. *Journal of Policy Research in Tourism, Leisure and Events.* doi: 10.1080/19407963.2016.1261146.

Séraphin, H., Gowreesunkar, V., & Ambaye, M., (2016). The Blakeley model applied to improving a tourist destination: The case of Haiti. *Journal of Destination Marketing Management.* doi: 10.1016/j.jdmm.2016.07.004.

Sharma, P., (2011). Country-of-origin effects in developed vs. Emerging markets: Exploring the contrasting roles of materialism and value-consciousness. *Journal of International Business Studies, 42*(2), 285–306.

Sharma, R., & Gupta, A., (2020). Pro-environmental behavior among tourists visiting national parks: Application of value-belief-norm theory in an emerging economy context. *Asia Pacific Journal of Tourism Research, 25*(8), 829–840.

Sok, P., O'Cass, A., & Miles, M. P., (2015). The performance advantages for SMEs of product innovation and marketing resources-capability complementary emerging economies. *Journal of Small Business Management.*

Stepchenkova, S., & Zhan, F., (2013). Visual destination images of Peru: Comparative content analysis of DMO and user-generated photography. *Tourism Management, 36*(2013), 590–601.

Tapper, R., (2001). Tourism and socio-economic development: UK tour operators' business approaches in the context of the new international agenda. *International Journal of Tourism Research, 3*, 351–366.

Tarlow, P., (2014). *Tourism Security: Strategies for Effectively Managing Travel Risk and Safety*. Elsevier.

Tranter, B., (2010). Environmental activists and non-active environmentalists in Australia. *Environmental Politics, 19*(3), 413–429.

Travalyst, (2022). *The Re-Emergence of Travel: The View of Travalyst*. Available from: https://travalyst.org/wp-content/uploads/2021/01/The-Re-emergence-of-Travel-Report-A-Travalyst-View.pdf (accessed on 20 January 2022).

Walters, G., & Mair, J., (2012). The effectiveness of post-disaster recovery marketing messages – the case of the 2009 Australian bushfires. *Journal of Travel & Tourism Marketing, 29*(1), 87–103.

Wang, C., Zhang, J., Xiao, X., Sun, F., Xia, M., & Shi, Q., (2020). Examining the dimensions and mechanisms of tourists' environmental behavior: A theory of planned behavior approach. *Journal of Cleaner Production*. 10.1016/j.jclepro.2020.123007.

Wei, Y. S., Samiee, S., & Lee, R. P., (2014). The influence of organic organizational cultures, market responsiveness, and product strategy on firm performance in an emerging market. *Journal of the Academy of Marketing Science, 42*(1), 49–70.

Wu, J., Font, X., & Liu, J., (2020). Tourists' pro-environmental behaviors: Moral obligation or disengagement? *Journal of Travel Research*. doi: 10.1177/0047287520910787.

Zarantonello, L., Schmitt, B. H., & Jedidi, K., (2014). How to advertise and build brand knowledge globally: Comparing television advertising appeals across developed and emerging economies. *Journal of Advertising Research, 54*(4), 420–434.

Zhang, Z., (2000). Developing a model of quality management methods and evaluating their effects on business performance. *Total Quality Management, 11*(1), 129–137.

CHAPTER 4

Postcolonial Branding and Self-Branding in Destination Marketing Strategy

HUGUES SÉRAPHIN,[1] XAVIER GUEGAN,[2] and VANESSA G. B. GOWREESUNKAR[3]

[1]*Tourism, Hospitality and Events Management, Oxford Brookes Business School, United Kingdom*

[2]*Department of History, The University of Winchester, Road Winchester Hampshire, SO22 4NR, UK*

[3]*Department of Cultural Tourism, Anant National University, Sanskardham, Bopal-Ghuma, Ahmedabad, Gujarat, India*

ABSTRACT

Branding plays a significant role in the performance of destinations. This management approach is particularly important for colonial destinations. The weight of the past plays a significant role in the branding strategy approach of destinations. Indeed, for some postcolonial destinations, the use of traumatic historical moments (which are part of their heritage) for marketing purposes, could also be considered as evidence of recovery from a traumatic past. As for non-postcolonial destinations, they are going the opposite way, as their branding strategy is based on neutrality.

Post-Disaster and Post-Conflict Tourism (An Updated 2nd Edition), Maximiliano E. Korstanje, Hugues Seraphin, and Vanessa G. B. Gowreesunkar (Eds.)

4.1 INTRODUCTION

The tourism industry as a modern phenomenon appeared in the mid-19th century with the modernization of transportation and communication and the resurgence of new empires, which connected this global economic activity to emerging tourists from different social classes. The psychological reasons that explain why people travel are oriented to the discovery of the other, in the terms of leisure, culture, business, health, and religion (Barrow, 2008; Draper, 2007; Davidson, 1998; Davidson & Maitland, 1997), as well as obtaining good experiences. Having said this, destination branding is esteemed as a long existing phenomenon. 'Brands and branding have existed for as long as it has been possible to trace artifacts of human existence' (Moore & Reid, 2008, p. 419). Hence it is important not to lose the sight of the fact of having a good knowledge of the history of corporate brands to comprehend the current strategy or an organization and provide guidance to management in terms of marketing and communication strategy for the future (Burghausen & Balmer, 2014a, b, 2015; Balmer & Burghausen, 2015). In the tourism industry, corporate brands used for communication with the tourism market (Walter & Mair, 2012), are important intangible assets that can have significant positive effects on the performance of a destination (Park, Eisingerich, Pol, & Park, 2013). As 'the present-centered nature of heritage is deeply associated with an increasing contemporary use of the past' (Park, 2014, p. 1), the purpose of this chapter is to investigate the conceptual relations and empirical dynamics between temporal concepts like history, colonial nostalgia – or its refusal, heritage, national identity, and organization branding. So doing, this chapter is going to rely on three different postcolonial destinations: Haiti as self-branding in postcolonial tourism; Western postcolonial branding and tourism of imperial past in India; and Algeria, which seems to have moved from colonial tourism to postcolonial rejection of tourism, and yet nostalgic stereotypes seem to endure.[1] This chapter is about the marketing use of 'heritage tourism,' defined by Park (2014) as using both

1 In this article, the use of the term 'postcolonial' refers to the temporality and thus period that follows the end of colonialism in these three countries, whereas 'postcolonial' is engaging with the ideological representation from both postcolonial and neo-colonial approaches.

material (tangible) and immaterial (intangible) remnants of the past. This conceptual and exploratory chapter that explores the concept of 'life of brands' that Da Silva Lopes (2007) defines as retracing the lives or trajectories of individual brand from their creation up to the present. In this chapter, the first section sets up the context of our study and explores the historical moments of each DMO chosen as a case study with an emphasis on the periods corresponding to the change of branding. In other words, we are going to explore the concept of 'life of brands' (Da Silva Lopes, 2007). The notion of 'life of brands' developed by Da Silva Lopes (2007) is related to the notion of 'storytelling,' which is fundamental in heritage tourism. There is therefore a clear link between heritage and marketing. The second section, the 'conceptual framework' is informed and shaped by the contextual framework. More specifically, we cover in our research frameworks the concepts of 'heritage,' 'branding' and 'collective memory' (nostalgia/national identity) that shape our approach.

4.2 HAITI, SELF-BRANDING IN POSTCOLONIAL TOURISM

The Americans occupied Haiti between 1915 and 1934 in order to prevent the Germans to take over banks and strategic areas in the country. This occupation positively impacted on Haiti (Saint-Louis, 2010; Charles, 1994) with the: improvement of infrastructures; construction of hospitals and vocational schools; reform of the organizations of the different public administrations; creation of the Haitian National Police and political stability. This context was favorable to the development of the tourism industry. The president at the time, Stenio Vincent (1939–1941) was the first to have highlighted tourism as a potential source of income for the country. As a result, the Haitian DMO created and its first logo designed in 1939 (Séraphin, 2014; Clammer, 2012).

The Haitian heritage is about religion. Voodoo is intimately linked to the culture and history of Haiti as the Haitians are 100% Voodooist and religion is the mainstay of Haitian society. Voodoo plays a key role in the Haitian survival of the slave trade and continues to provide the locals with a common identity and connection with their past. Voodoo is for Haiti, a shared understanding and acceptance of a common identity (Séraphin & Nolan, 2014). The black face of the mambo (female high priest in the voodoo religion in Haiti) in the first logo of the DMO is reminiscent of this idea. This logo is also reminiscent of the fact that Haiti was a rich colony, with agriculture (and mainly the growth of coffee), playing an important role in its prosperity (Destin, 2014; Roc, 2008). The Haitian DMO gives importance to heritage in its branding strategy. The changes in the political context of the country pushed the DMO to turn toward the dominant image and largest tourist market of 'sun-and-sand' (Nelson & Torres, 2010). This denotes a will from the destination to get away from its historical colonial image to a more modern society. This strategy goes against Bancel (2005) research on ethnic tourism and colonialism in which the logo is seen as a will from the destination to brand itself as a natural, authentic, welcoming destination with strong and stable historical background and landmarks untouched by globalization. That said, Bancel (2005) acknowledged the limitations of ethnic tourism in postcolonial context as it could negatively impact on tourism as it could put face to face the tourists with the post-colonial reality (Wilton, 2005). Other DMOs in the Caribbean seem to be moving away from a marketing strategy based on heritage toward a more mainstream and neutral approach.

4.3 WESTERN AND NATIONAL POSTCOLONIAL BRANDING AND TOURISM OF IMPERIAL PAST IN INDIA

From the mid-19th century, the British developed the tourist industry within their Indian Empire. It was mainly created for the British administration, civil servants, soldiers, entrepreneurs, and businessmen, but also for the small group of European visitors who came to visit their relatives. India since its independence has continued to generate a growing tourist industry. It has developed marketing strategies that are similar with other countries in the region and 15 of its states have their own tourism logo, which is usually linked with the local wildlife and environment, and sometimes iconic heritage. Since 2002 India's Ministry of Tourism has developed logos and brandings around the name 'Incredible India' as part of an international marketing campaign to promote tourism in India to an audience of global appeal (Official website of India's Ministry of Tourism: http://incredibleindia.org/). On the other hand, the tour operators that promote India tourist circuits within their brochures do not really engage with Indian self-representation of its culture, sites, and nation, but instead often continue to use tropes, images, and themes from the colonial time. Amongst the well-established type of guides that are promoting and encouraging the tourist to go traveling around the world are the *Lonely Planet* guidebooks (and today, also the website), which is the largest travel guide book publisher in the world (Fildes, 2007). Created in the early 1970s, *The Lonely Planet* guidebook series soon after expanded in Asia with the India guidebook that was first published in 1981.

4.4 ALGERIA, FROM COLONIAL TOURISM TO POSTCOLONIAL REJECTION OF TOURISM

In 1849 the Algiers French Municipal Council cited the city's climate and striking setting as reasons to believe that two recent projects – a new theater and a promenade through the hills above the city – could provide recreation not only for the local population but also for a growing number of seasonal visitors. Thus did a tourist infrastructure emerge, and with it a surge in the number of visitors who made the years until 1914 an era of continuous touristic development. A stay in Algiers figured in the itinerary of many European travelers, who went to the French Algerian *départements* to experience different kind sites, such as the lands of antiquity, the mountain and desert landscapes which were the other main sources for their impressions of the country beyond the 'modern' French Algerian cities. As for British India, the French developed in Algeria a tourist market around souvenirs, and Rebecca DeRoo observes that at the turn of the century 'postcards became mass media of communication and collectible objects for the first time in French history' and that a 'significant portion of the millions of postcards produced yearly in France displayed Algerian tourist sites and ethnic types' (DeRoo, 2002, p. 159). The colonial consciousness that connected the metropole to the periphery, which institutionalized the presence of France and Britain, respectively in Algeria and India highlighted the similarity between both imperial systems demonstrating the formation of an imagined imperial community via trans-national and trans-imperial agents – such as governmental actors, modern urbanization, markets for images, technical advance, public, and private companies, communication, and networks of suppliers, learned societies, and tourist development and tours (Guégan, 2015). However, at the difference of India, the Western tourist branding of Algeria did not follow the same postcolonial trajectory. Indeed, since the 1980s – and even more during the Civil War of the 1990s – the country has closed itself to international tourism, although in the last decade, it has started to try to change its attitude towards this industry.

4.5 POSTCOLONIAL BRANDING, HERITAGE, AND COLLECTIVE MEMORY

Park's research on culture and heritage tourism tells us much about how we can classify heritage building. He finds three main groups covering a wide range of areas (including culture; history; politics; religions, communications, etc.), that can be classified into three main groups: built heritage (forts, relics, etc.); scientific heritage (plants, birds, natural habitats, etc.); and cultural heritage (fine art, customs, languages, etc.). Nostalgia plays an important role in the rising popularity of heritage as the latest appears as a secure and stable platform, hence its democratization. Heritage may be digested following a re-enactment of the past. To put this slightly in other terms, the epistemology of heritage derives from a symbolization of the past, which re-enact with locals and tourists in the present. As a result, the tourism industry has often been blamed for the commodification of heritage sold to tourists as products and services. Because of the commodification of heritage, authenticity becomes an issue. The form of tourism based on heritage is even referred to as 'staged authenticity' and yet authenticity is considered as the most important criterion for the development of heritage tourism. 'Identity' is another key word associated with 'heritage.' Heritage in this instance enables a group to feel and remain connected to their ancestral roots. As for heritage marketing, the heart of this chapter, Park (2014) explains that this type of marketing is all about the use and very often the overuse of cultural symbols, historical values, sacred icons, images, and stories in order to develop an emotional connection between heritage and (potential) tourists. The collective memory adapts the past to the eyes of the present (Zelinzer, 2008, cited in Volcic et al., 2014).

As this backdrop, Volcic (2008) acknowledges that the media plays a leading role as: (i) archive of collective memories; (ii) Constructor of collective memory; (iii) agents to branding traumatic past; and (iv) adjust representation of past events to the contemporary political and social mentality and sensibility of the present. As a logo is a communication media (Park, 2014; Park et al., 2013; Morgan, Pritchard, & Pride, 2011;

Marti, 2008; Saint-Hilaire, 2005) we are arguing that it plays the same role as Volcic's nomenclature of journalists. Dean MacCannell explains tourists are on a quest for 'authenticity' in the 'other,' and John Urry states that there is no authentic tourism and that tourists merely seek contrasts, aware of the inauthenticity of sights and sounds (MacCannell, 1976, p. 94; Urry, 1990). In any case, the authenticity of the experience is really of concern for the collective gaze behind postcolonial branding (Farr & Guégan, 2013). This dimension does seem to have been taken into account by DMOs. Indeed, the results of a research carried out by the UK Tour Operator *Family Break Finder* (2016 [Online]) revealed that out of 193 countries DMO slogan surveyed none of them have any keywords relating to heritage, identity, history, etc. The words used in the slogans are more marketing orientated and also more mainstream.

If we believe Séraphin, Ambaye, Gowreesunkar, & Bonnardel (2016) a good logo design should take into account the context, history, identity, and essence of a destination. On that basis, we can assume that any postcolonial destination branding itself overlooking its history colonial heritage would not tell the full narrative of the destination and would undoubtedly fail from a marketing point of view.

Colonial empires like France and England are using strong symbols of their heritage in their branding strategy. We can come to the conclusion that former colonies and former colonial empires are using opposing branding strategies. At this stage of our research, it is legitimate to come to an interim conclusion that heritage can be either a blessing or a curse.

4.6 BRANDING, SELF-BRANDING, TRAVELING, AND IDENTITY

The rapport between Britain, France, and the United States and their respective empires in postcolonial tourism is twofold: on the one hand there is the general politics and phenomena of certain actions and policies coming from the governments of India, Algeria, and Haiti; and on the other there is the particular, with individuals and tour operators from the old metropoles who take part in the shaping of those global patterns of memory and identity by creating common tropes and cultural ideologies that still link, economically and culturally, the 'old' and 'new' countries. John M. MacKenzie reinforces the point that empires' reassurance was created mainly through constructed fantasies (MacKenzie, 2013). Conceived of as ruling the world, their fantasy was essentially governmental. Imperial guidebooks became a significant part of the international and colonial travel scene in the 19th and twentieth centuries, and postcolonial branding has somehow continued this tradition, perhaps at the exception of Algeria. The imperial mindset with its postcolonial alter ego approaches the economies of the territories that they cover, the environments, natural histories and historic monuments of such apparently remote and exotic places, and above all to the social mores of colonial settlers and expatriates as well as to relationships with and attitudes towards still perceived 'indigenous' people. The insights that they provide into imperial/postcolonial touring, as well as the manner in which they reveal the ways (mainly bourgeois) in which Western tourists envision the world even up to the twenty-first century. The significance is that the branding and touristic approach of Haiti, India, and Algeria has still so much to do with 'Americanness,' 'Britishness' and 'Frenchness.'

Experiencing a postcolonial world thus brings the views and perceptions from the old metropoles into the construction of memory and the evolution of representation of that world. However, for the Haitians, Indians, and Algerians, their relationship to that mythical branding is actually not often interfering with their direct life, and their perception of and identification with the Western tourists is very different from these tourists' postcolonial visions. Anna Bocking-Welch highlights that Gavin Young in his 1991 book "*In Search of Conrad*" described his travels in Southeast Asia as a ghost hunt 'among the echoes and shadows of a flickering past' (Bocking-Welch, 2013). Ghost-hunting as the travel narrative behind branding since the 1960s is the precise appeal of the imperial past to travelers and tourists that has shifted over time. As discussed in earlier chapters, the education of tourists is the touchstone of post-disaster marketing and consumption. This is a preliminary approach that merits further discussion.

4.7 POSTCOLONIAL DESTINATIONS AND BRANDING IN THE PANDEMIC CONTEXT

As the world is still navigating within an on-going COVID-19 pandemic context, tourism destinations are doing their best to catch up with their losses; international and domestic tourism suffered a dramatic downturn which surpassed the losses suffered by this sector of the economy following earlier major events, including the '9/11' terrorist attacks (2001); SARS (2003); the Indian Ocean Tsunami (2005); MERS (2012); swine flu

(2009); Ebola (2014), and other crises in the twenty-first century (Zhan et al., 2020; Higgins-Desbiolles, 2020). Undoubtedly, several recovery strategies have been proposed by the United Nations World Tourism Organization (UNWTO) and destinations are generally recovering. However, in the context of postcolonial destinations, the pandemic has come both as a blessing and a blight. According to Séraphin & Gowreesunkar (2016); and Séraphin et al. (2017), postcolonial destinations generally carry a negative image, and their tourism performances are not too desirable. Drawing from the Caribbean islands, the authors explain that the region's poor reputation has been a major reason for under visitation since the past three decades. For instance, Haiti is branded as an insecure destination and as a place where the worst is always likely to happen and arguably, tourists are reluctant to travel to such place due to its image of 'diseased destination' (Séraphin et al., 2017; Maximiliano et al., 2020). In contrast, postcolonial successful destinations like Mauritius and Seychelles have continued to attract tourists (mainly European) as their branding strategies are mainly based on history and heritage. In the case of Mauritius, French tourists are interested in visiting the island as they are curious to discover the legacy of Mahé de Labourdonnais, the First French Governor of the island. Likewise, Lonely Planet describes Port Louis, as the island's capital and largest city with flashes of French and British culture (https://www.lonelyplanet.com/mauritius/port-louis). The oldest post-office in the city and the Trou Fanfaron police station are recognized as national heritage of Mauritius, while the cathedral built by Robert Farquhar in 1933 is another important pull factors in tourism. In the case of Seychelles, the branding strategy is centered around colonial imaginaries (Kothari & Wilkinson, 2010). Island imaginary has become an important cultural and economic resource for the tourism industry. The national museum of Victoria in the capital city Victoria is another of colonial legacy used as pull factors in tourism. These studies demonstrate how the use of French legacy is a postcolonial phenomenon in tourism and history has been converted into strength by re-staging colonial heritage in the interest of both islands' tourism industry. In these cases, the heritage branding is seen as an efficient instrument for postcolonial destinations to gain further attention in the international market (Maximiliano et al., 2020).

The arrival of COVID-19 has changed the whole scenario for struggling postcolonial destinations. Postcolonial destinations heavily dependent on tourism, have shown sign of concern for their economy regarding the deep and prolonged downturn in the travel and tourism sector has made them very concerned about their finances (UNCTAD, 2020). For example, the OECD policy responses to COVID-19 provides significant examples of postcolonial economies which collapsed, and which are desperately looking for bold government action and adequate international support (OECD Report, 2021). Already before the COVID-19 pandemic hit, postcolonial destinations were vulnerable to external shocks due to their unique structural characteristics, and they have been struggling with rising public and private debt levels. Their remoteness determines that, as a group, they are less than one third as well connected as other developing countries (OECD, 2018), and this reduces their access to international markets and competitiveness. These economic vulnerabilities are inter-related and reinforced by the climate and environmental challenges, such as increasingly frequent extreme weather events, rising sea levels, ocean acidification, loss of ocean oxygen and ecosystem degradation.

Although postcolonial destinations are extremely vulnerable to natural, social, environmental, political, and cultural risks, the industry have developed a resilience vis a vis the pandemic (Rahman et al., 2021). The diseased destination image is now seen as desired destination; post-pandemic tourism is now about less crowded, less visited and insular destinations. Frustrated travelers of the 2020 lock-down are no longer attracted to 'Americanness,' 'Britishness' and 'Frenchness,' as these places have been labeled as COVID-breeding countries. Branding a destination using a resilient and restoration strategy is not new (Pereira et al., 2021), but in the specific case of the COVID-19 pandemic, it is taking a somewhat different turn. For instance, while the world is still lamenting on its losses ($1 trillion in global tourism economies and more than 100 million jobs as reported by the United Nations Policy Brief, 2020), researchers and observers have started seeing the bright sides of the pandemic (Gowreesunkar et al., 2021; Subadra et al., 2021; Mugizi et al., 2021). Vulnerabilities of postcolonial destinations are now increasingly being seen as opportunities. According to them, the pandemic is giving a breathing space to tourism resources. This is also confirmed in a number of world reviews (*see* Euronews, 2020; Hindustan Times, 2020; Gowreesunkar et al., 2022). For instance, established postcolonial tourism destinations like Seychelles and Mauritius, which have somehow suffered from overtourism due to pressure on their tourism resources are now recovering (Gowreesunkar & Rycha, 2015). The lack of tourists amid

COVID-19 has allowed the destinations to rethink their tourism strategies, along with how the issue of overtourism can be managed effectively. Likewise, in India, after many decades, pollution levels dropped, and several species of animals, birds, and insects have staged a comeback. The pandemic has also reinforced the position of Séraphin & Gowreesunkar (2019) on the double facet of the industry: tourism has a Janus-faced character. For instance, within the space of months, the framing of the global tourism system moved from overtourism to non-tourism (Gössling et al., 2021). Likewise, the label of 'diseased destination' was converted to 'desired destination,' as less desirable postcolonial islands are less crowded. As the world is still navigating in between the '*new normal*' and the '*next normal*' of an ongoing pandemic, postcolonial destination are certainly investing into new branding strategies. For instance, Uganda, a postcolonial destination which did not enjoy a great tourism reputation, is now gaining popularity as it portrays its destinations as less crowded with fewer locals. The destination's main strategy is based on health and safety of tourists, and Uganda is now seen as a destination capable of dealing with diseases and pandemics. International tourism has been encouraged by a re-branding strategy.

The world has been transformed by the COVID-19 crisis, and impacts are expected to be more severe than the 2008 financial crisis and truly global in reach (OECD, 2020). Seen differently, the COVID-19 has offered an opportunity to postcolonial destinations to re-invent their branding strategies based on a 'new normal' and postcolonial destinations are seen as desired destinations. The Corona crisis has brought more innovation and positive impact through various untapped opportunities. Even if the negative impact of the pandemic is inevitable, at the same time, it opens a window of opportunity to invest in new models of tourism activities in postcolonial destinations previously struggling for a position in the global tourism map.

4.8 CONCLUSION

The purpose of this conceptual study was to advance understanding on the importance of heritage in branding strategies of postcolonial, destinations. Exploratory in nature, the study theoretically concludes that the use of traumatic historic moments (heritage) for marketing purpose does not lead to a potential trivialization and heritage detachment. On the contrary, they could be considered as a symbol of recovery from trauma by the locals. That said, it seems that postcolonial destinations and non-postcolonial destinations are going for neutrality, mainstream strategy. Theory and the reality of life in this chapter are not going to the same direction.

KEYWORDS

- **branding**
- **neutrality**
- **nostalgia**
- **postcolonial destination**
- **trauma**
- **weight of the past**

REFERENCES

Balmer, J. M. T., & Burghausen, M., (2015). Introducing organizational heritage: Linking corporate heritage, organizational identity, and organizational memory. *Journal of Brand Management, 22*(5), 385–411.

Bancel, N., (2005). Ethnic tourism: A symbolic reconquest? In: Blanchard, P., & Bancel, N., (eds.), *Traces and Colonial Memories in France*. Paris: Otherwise.

Barrow, S., (2008). Edutourism – there is no darkness but ignorance. *Tourism Society Journal, 137*, 2, 3.

Bocking-Welch, A., (2013). Ghost hunting: Amateur film and travel at the end of empire. In: Martin, F., & Xavier, G., (eds.), *The British Abroad Since the Eighteenth Century: Experiencing Imperialism* (Vol. 2). Basingstoke: Palgrave Macmillan.

Burghausen, M., & Balmer, J. M., (2014). Corporate heritage identity management and the multi-modal implementation of a corporate heritage identity. *Journal of Business Research, 67*(11), 2311–2323.

Burghausen, M., & Balmer, J. M., (2014). Repertoires of the corporate past: Explanation and framework. Introducing an integrated and dynamic perspective. *Corporate Communications: An International Journal, 19*(4), 384–402.

Burghausen, M., & Balmer, J. M., (2015). Corporate heritage identity stewardship: A corporate marketing perspective. *European Journal of Marketing, 49*(1, 2), 22–61.

Charles, E., (1994). *Political Power in Haiti from 1957 to Our Days*. Paris: Karthala.

Clammer, P., (2012). *Haiti*. Guildford: Bradt.

Da Silva, L. T., & Casson, M., (2007). Entrepreneurship and the development of global brands. *Business History Review, 81*, 651–680.

Davidson, R., & Maitland, R., (1997). *Tourism Destination*. Hodder and Stoughton.

Davidson, R., (1998). *Travel and Tourism in Europe*. London: Longman.

DeRoo, R. J., (2002). Colonial collecting, French women and Algerian *cartes postales*. In: Gary, D. S., & Eleanor, M. H., (eds.), *Colonialist Photography, Imag(in)ing Race and Place*. London: Routledge.

Draper, R., (2007). Tennis' contribution to the British economy. *The Tourism Society Journal, 132*, 13.

Farr, M., & Guégan, X., (2013). *The British Abroad Since the Eighteenth Century: Travelers and Tourists* (Vol. 1). Basingstoke: Palgrave Macmillan.

Farr, M., & Guégan, X., (2013). *The British Abroad Since the Eighteenth Century: Experiencing Imperialism* (Vol. 1). Basingstoke: Palgrave Macmillan.

Fildes, N., (2007). BBC Gives Lonely Planet Guides a Home in First Major Acquisition. The Independent.

Gaitree, G., & Ian, R., (2015). A study on the impacts of dolphin watching as a tourism activity: Western Mauritius as case study. *International Journal of Trade, Economics and Finance, 6*(1), 67.

Gössling, S., Scott, D., & Michael, H., (2020). Pandemics, tourism and global change: A rapid assessment of COVID-19. *Journal of Sustainable Tourism, 29*(1), 1–20. doi: 10.1080/09669582.2020.1758708.

Gowreesunkar, V. G. B., Maingi, S. W., Roy, H., & Micera, R., (2021). Introduction: Destination management solutions post pandemic: A need of the hour. Pandemic and the accommodation sector in Sardinia, Italy: Impacts and response actions. In: Gowreesunkar, V. G., Maingi, S. W., Roy, H., & Micera, R., (eds.), *Tourism Destination Management in a Post-Pandemic Context (Tourism Security-Safety and Post Conflict Destinations)* (pp. 49–65). Emerald Publishing Limited, Bingley. https://doi.org/10.1108/978-1-80071-511-020211004.

Gowreesunkar, V., & Reddy, M., (2020). Urbanism and over tourism: Impacts and implications for the city of Hyderabad. *Routledge Handbook of Tourism Cities*. Emerald Publishing, UK.

Gowreesunkar, V., Maingi, S., Roy, H., & Micera, R., (2021). *Tourism Destination Management in a Post Pandemic Context: Global Issues and Destination Management Solutions*. Emerald Publishing Limited, First Edition, UK.

Guégan, X., (2015). Transmissible sites: Monuments, memorials and their visibility on the metropole and periphery. In: Geppert, D., & Lorenz, M. F., (eds.), *Sites of Imperial Memory: Commemorating Colonial Rule in the Nineteenth and Twentieth Centuries*. Manchester: Manchester University Press.

Higgins-Desbiolles, F., (2020). Socializing tourism for social and ecological justice after COVID-19. *Tourism Geographies*. Online. doi: 10.1080/14616688.2020.1757748.

Kothari, U., & Wilkinson, R., (2010). Colonial imaginaries and postcolonial transformations: Exiles, bases, beaches. *Third World Quarterly, 31*(8), 1395–1412.

MacCannell, D., (1976). *The Tourist: A New Theory of the Leisure Class*. New York: Schocken.

MacKenzie, J. M., (2013). Empire travel guides and the imperial mindset from the mid-nineteenth to the mid-twentieth centuries. In: Martin, F., & Xavier, G., (eds.),

The British Abroad Since the Eighteenth Century: Experiencing Imperialism (Vol. 2). Basingstoke: Palgrave Macmillan.

Marti, B. E., (2008). Cruise line logo recognition. *Journal of Travel and Tourism Marketing, 18*(1), 25–31.

Maximiliano, M., Séraphin, H., & Gowreesunkar, V., (2020). *Postcolonial, Conflict and Disaster Destinations Marketing Strategy: Towards a New Approach to Tourism Management*. Book in press at Apple Academic Press, a division of Taylor & Francis Group, CRC Press.

Morgan, N., Pritchard, A., & Pride, R., (2013). *Destination Brands: Managing Place Reputation* (3rd edn.). London: Butterworth-Heinemann.

Mugizi, F., Ayorekire, J., & Obua, J., (2021). Uganda tourism sector COVID-19 response, recovery and sustainability strategies: Lessons from previous disease outbreaks. In: Gowreesunkar, V. G., Maingi, S. W., Roy, H., & Micera, R., (eds.), *Tourism Destination Management in a Post-Pandemic Context: Global Issues and Destination Management Solutions.*

Nelson, V., & Torres, R. M., (2010). Conceptualizing Caribbean tourism through hybridity: The Grenadian tour product. *Journal of Tourism Research ARA, 92*(103), 92–103.

OECD, (2020). *COVID-19 Survey – Main Findings*. OECD Publishing, Paris, https://one.oecd.org/document/DCD/DAC/STAT(2020)35/en/pdf (accessed on 20 January 2022).

OECD, (2020). *COVID-19: Understanding Health Risks in Small Island Developing States (SIDS) DCD/DAC/2020(36)*, OECD (not publicly available).

OECD, (2021). *COVID-19 Pandemic: Towards a Blue Recovery in Small Island Developing States*. OECD Policy Responses to Corona Virus, OECD.

Park, H. Y., (2014). *Heritage Tourism*. Abingdon: Routledge.

Pereira, A., Frias, C., & Jerónimo, A. P., (2021). Falling in love again: Brand love and promotion of tourist destinations during the COVID-19 pandemic. In: Seabra, C., Paiva, O., Silva, C., & Abrantes, J. L., (eds.), *Pandemics and Travel* (*Tourism Security-Safety and Post Conflict Destinations*) (pp. 227–241). Emerald Publishing Limited, Bingley.

Roc, N., (2008). Haiti-environment: From the 'pearl of the Antilles' to desolation. *FRIDE*, 7.

Saint-Hilaire, B., (2005). *The Brand: Create and Develop Your Identity*. Paris: Vuibert.

Séraphin, H., & Gowreesunkar, V., (2017). What marketing strategy for destinations with a negative image? *Worldwide Hospitality and Tourism Themes, 9*(5), 496–503. https://doi.org/10.1108/WHATT-07-2017-0035.

Séraphin, H., & Nolan, E., (2014). Voodoo in Haiti: A religious ceremony at the service of the 'houngan' called 'tourism.' In: Frost, W., & Laing, J., (eds.), *Rituals and Traditional Events in the Modern World*. New-York: Routledge.

Séraphin, H., (2014). *Tourism: Opening for the People of Toussaint?* Paris: Publibook.

Séraphin, H., Bonnardel, V., & Gowreesunkar, V. G. B., (2020). Redesigning Haiti's destination marketing organization logo: Identity, aesthetics and communication implications. *Journal of Destination Marketing and Management, 15*, 100393.

Séraphin, H., Gowreesunkar, V. G. B., & Blakeley, K., (2017). Conceptual analysis and application of a lifelong learning in tourism approach to maximize the performance of Haiti as a tourist destination. *Lifelong Learning for Tourism: An International Perspective.*

Subadra, I. N., (2021). Destination management solution post COVID-19: Best practice from Bali – A world cultural tourism destination. In: Gowreesunkar, V. G., Maingi, S. W., Roy, H., & Micera, R., (eds.), *Tourism Destination Management in a Post-Pandemic Context (Tourism Security-Safety and Post Conflict Destinations)* (pp. 25–36). Emerald Publishing Limited, Bingley. https://doi.org/10.1108/978-1-80071-511-020211024.

Urry, J., (1990). *The Tourist Gaze: Leisure and Travel in Contemporary Societies*. London: Sage.

Volcic, Z., Erjavec, K., & Peak, M., (2014). Branding post-war Sarajevo. *Journalism Studies, 15*(6), 726–742.

Zhan, L., Zheng, X., Morrison, A. M., Liang, H., & Coca-Stefaniak, J. A., (2020). A risk perception scale for travel to a crisis epicenter: Visiting Wuhan after COVID-19. *Current Issues in Tourism*.

CHAPTER 5

Global Branding Strategy and Framework

HUGUES SÉRAPHIN,[1] KAREN CRIPPS,[2] and SIMON M. SMITH[3]

[1]*Department of Marketing, Event Management and Project Management, The University of Winchester, Winchester, Hampshire, SO22 4NR, UK*

[2]*Department of Responsible Management and Leadership, University of Winchester Business School, Winchester, Hampshire, SO22 4NR, UK*

[3]*Oxford Brookes Business School, Oxford Brookes University, Oxford, OX3 0BP, UK*

ABSTRACT

Although tourism marketing and management have gained firm positions over the recent years, no less true seems to be that the rules of marketing have been put upside down just after the pandemic. The DMO model has been proven to be efficient in recycling destinations in the context of crisis and chaos. This chapter delves into the world of national identities and heritage as key factors of national cohesion and unity. Having said this, the DMO model helps readers to expand their understanding of tourist preferences as well as the narratives emanating from the core of nations and sold as national heritage.

Post-Disaster and Post-Conflict Tourism (An Updated 2nd Edition), Maximiliano E. Korstanje, Hugues Seraphin, and Vanessa G. B. Gowreesunkar (Eds.)

5.1 INTRODUCTION

Before COVID-19, tourism was a dynamic, fast-growing, and an important global key driver of socio-economic growth, and it is not confined to any one region (Ashley et al., 2007; Gopalan, 2013; Malik et al., 2010; Thano, 2015). Currently, Europe is still the world's largest source region, generating half of the world's international arrivals, followed by Asia and the Pacific (24%), the Americas (17%), the Middle East (3%), and Africa (3%) (UNWTO, 2016). Moreover, estimates from the UNWTO Tourism Towards 2030 suggest that the number of international tourist arrivals worldwide is expected to increase by an average of 3.3% a year over the period 2010 to 2030. Of course, these figures can expect to be adjusted because of COVID-19's impact between 2020 and until we reach normality within travel and tourism. Therefore, although international tourist arrivals worldwide are expected to reach 1.4 billion by 2020 and 1.8 billion by the year 2030, we can perhaps concentrate more on that latter figure. In 2030, 57% of international arrivals are expected to be in emerging economy destinations (versus 30% in 1980) and 43% in advanced economy destinations (versus 70% in 1980) (UNWTO, 2016).

In that respect, the process of branding is an important aspect of destination management that contributes to the competitiveness of the destination (Galì et al., 2016). This chapter is going to provide a global overview of the branding strategy adopted by destinations. More specifically, we are going to focus on the place of heritage and identity in their branding strategy. Our research question is therefore as follows: How are national identities and heritage reflected in a DMO's slogan? Because 'heritage plays a fundamental role in enhancing the identity of a nation or region and is a major vehicle for expressing national and regional cohesion and unity' (Park, 2014, p. 108), from a theoretical and managerial point of view, it is important to understand whether or not the branding strategy adopted conveys the narrative of the destination. From a methodological point of view, this chapter focuses on a qualitative perspective. The progressive structure and the inductive approach adopted within the chapter will lead to us providing a fairly straight answer to the research question.

5.2 THE AMERICAS (CARIBBEAN DESTINATIONS)

The Americas include North America, South America, and Central America (including the Caribbean). The Americas have the second most improved travel and tourism competitiveness index (TTCI[1]) performance at the aggregate level, just behind Asia-Pacific. International tourist arrivals have boomed, growing from 170 million in 2013 to over 201 million in 2015. The majority of the countries in the region rely on rich natural resources and good service infrastructure to appeal to tourists. While North and Central American nations outperform South American ones on infrastructure, the opposite is true for cultural resources. Numerous South American nations are taking advantage of their rich heritage to create a strong tourism value proposition consisting of natural resorts, entertainment, and culture. On the other hand, Central American and Caribbean countries continue to rely too excessively on their natural resources and have not made much progress in developing other tourism segments (WEF, 2017 [Online]).

Because of the Caribbean destinations are very different from the other destinations of the Americas, mainly in terms of size, carrying capacity in terms of tourists, facilities, culture, etc., we have decided to overlook the other components of the Americas. This approach is going to provide more accurate results. Moreover, the destinations in the Caribbean are quite similar at many levels:

- They belong to the Creole society or black culture (Reyes-Santos, 2013; Cashmore, 2006; Green, 1972).
- They have a common ground of political instability due to the fact that 'the Caribbean was ruled by some of the most infamous dictators of the Western Hemisphere' (Bucheli, 2008).
- The islands face similar issues like drug fight, preservation of heritage, natural disaster management, immigration management, etc. (Dubesset, 2013).

In their branding strategy, the Caribbean destinations are mainly focusing: (i) on the authenticity and the quality of the experience that the visitors can get. A large proportion of destinations are also adopting;

---

1. The Travel and Tourism Competitiveness Index (TTCI) measures 'the set of factors and policies that enable the sustainable development of the travel and tourism sector, which in turn, contributes to the development and competitiveness of a country. It comprises 4 sub-indexes, 14 pillars, and 90 individual indicators, distributed among the different pillars (WEF, 2017 [Online]).

and (ii) a neutral tone in their slogans. Our findings contrast with the fact that Caribbean countries continue to rely too excessively on their natural resources (WEF, 2017 [Online]). There is here a discrepancy between the actual strategy implemented by the DMO and the branding strategy of the destination. That said, this trend toward neutrality is also reflected in the DMOs logo. If we consider, for instance, the Haitian DMO, their original logo designed in 1939 it was quite representative of the history and culture of the destination. In the process of changing the negative image of the destination, the DMO adopted in 2012 a more commercial/neutral logo.

5.3 AFRICA

Sub-Saharan Africa remains on aggregate, the region where Travel and Tourism competitiveness is the least developed. Although regional performance has increased, it has improved less compared to other parts of the world. Southern Africa remains the strongest sub-region, followed by Eastern Africa and then Western Africa. Tourism in the region is mainly driven by natural tourism. In several African countries, there are numerous cultural sites and intangible expressions that could be better leveraged and combined with the rich natural capital available (WEF, 2017 [Online]). The branding strategy trend in Africa seems to go toward neutrality This could be explained by the fact that the African continent wants to move away from its painful past related to slavery. The continent is trying to reinvent itself.

Among the various logos of this area, Egypt is particularly interesting. In an attempt to expand its touristic activity beyond its history (made up of pyramids and ancient ruins), Egypt has introduced a new identity through a new logo that suggests another side of Egypt, that of the beach (the Red Sea, the Sinai Peninsula, and the Gulf of Aqaba) (Avraham, 2016).

5.4 ASIA/OCEANIA

Asia and the Pacific area continue to be among the fastest growing tourism regions in the world. In 2015, international tourist arrivals in the region grew by 6%, reaching 279 million more than the only 110 million in 2000. By 2030, according to UNWTO Tourism Towards 2030, growth will be more and more pronounced. Asia and the Pacific, and especially China, is rapidly becoming the main source of many destinations, i.e., within the region. About 80% of international tourism in Asia is indeed intra-regional (UNWTO/GTERC, 2016).

Despite this growth, this region is not yet mature from the point of view of the destination branding strategy. In fact, most slogans are considered neutral. There is also no reference to historical aspects. The logo of Singapore reflects the overall trend toward neutrality, the purpose being probably to be more visitor-centric and connecting directly with consumers.

5.5 EUROPE

Europe is the region with the strongest overall travel and tourism competitiveness performance. It boasts six of the 10 most competitive countries in the travel and tourism sector and attracted 620 million of the 1.2 billion international visitors in 2016. To date, strong health and hygiene conditions, cultural richness and safety are common travel and tourism competitiveness advantages across the majority of European countries (WEF, 2017 [Online]). The category more present is the "neutral," followed by "quality of life." Europe is the geographical area in which we found the most DMO strategy based on history. For this area, history is a key pull factor. In France for instance, the Eifel Tower is the most visited attraction. Taking Spain as another example, it is among the best examples of modern national branding, with its easy recognizable Mirò's sun designed in 1983, which symbolizes the modernization of the country (Gilmore, 2002).

5.6 TREND PER AREA OF THE WORLD

Regarding the distribution of the category (Table 5.1) it's interesting to note that "natural features" (scientific heritage) category is most represented in Africa, while the "quality of life" (cultural heritage) is in Europe. In the same way, "history" (built and cultural heritage) is especially in Europe and Africa while "authenticity of experience" (cultural heritage) is in the Caribbean. The overall results of our research show that areas of the world base their marketing strategy on their endowments (either natural or built heritage).

TABLE 5.1 Branding Strategy of the DMO

	Europe	Asia	Caribbean	Africa	Total
Natural features	4	2	3	4	13
Neutral	22	36	5	13	76
Quality of life	8	5	1	3	17
Geographical features	4	2	0	0	6
Authenticity of experience	4	4	5	1	14
History	4	0	0	3	7

Basically, there are two main streams for a logo and slogan design. Either a neutral position is adopted or a narrative position (scientific, cultural, and built heritage). Based on Séraphin et al. (2016), a good logo design should adopt an ambidextrous approach, i.e., be able to tell the narrative of the destination (scientific, cultural, and built heritage), be a good marketing and communication tool, and be esthetically pleasant. This ambidextrous approach can be important to a destination. In simple terms, Organizational Ambidexterity is all about embracing two opposites, namely *exploitative* and *explorative* approaches (Smith, 2016, 2017; Stokes et al., 2015). The theory here is applied to destinations rather than individual organizations. This is a complex positioning but allows this theoretical lens to offer different considerations and approaches within destinations. In essence, a destination needs to manage two potentially opposing and contradicting forces. On the one hand, there are *exploitative* approaches, i.e., generally focused on what is known (underpinned by convergent thinking) and centered on existing customers/tourists and markets. The polar opposite is explorative approaches, i.e., generally focused on moving beyond that existing knowledge into new knowledge, customers/tourists and markets through embracing aspects like innovation, experimentation, flexibility, and divergent thinking. This approach contributes to the sustainable development of an organization (Mihalache & Mihalache, 2016).

Organizational Ambidexterity is a theory pertaining to paradox (Smith, 2016). Those polar opposites previously mentioned are potentially paradoxical within a destination and these tensions could ultimately lead to poor decision making within a destination, i.e., by following one extreme or the other to the detriment of the polar opposite. Smith (2016) proposes a need to maximize the potential of the paradox and 'embrace' it in all its complexities and contradictions. In essence, if these tensions can be managed effectively, it could lead to greater benefits for the destination.

As an example, one particular paradoxical tension is the need for affordability for a tourist to travel to a destination versus highly differentiated and unique experiences. The former can be cost efficiency related and connected to *exploitative* approaches. The latter can require investment and innovation and relates to *explorative* approaches. If a destination can lead in both sets of approaches, this could be crucial to success. But of course, it is rife with the problem of being paradoxical.

Netflix (although not a tourist destination) is a great example of embracing the extremes. They use globalization, technology, and economies of scale to maximize *exploitative* approaches and then inject a large-scale set of costly, innovative, and therefore *explorative* practices through the creation of unique film and TV. A monument to their success is the inability of competitors to wrestle away market share from Netflix. This is because (put in fairly crude terms) they keep the subscription rates for customers low enough to resist low-cost entrants and constantly add innovation to fight off companies or new entrants attempting to offer high differentiation. If destinations can manage such ambidextrous practices, this could set them apart from the competition. The importance of such dynamic approaches is perhaps amplified by the arrival, devastating impact and subsequent normalization in everyday life of COVID-19, which has brought so many other complexities into tourism practices.

This chapter examines the tourism slogans of 139 destinations around the world. Findings have revealed that neutrality (no use of scientific, cultural, and built heritage) is by far the mainstream strategy adopted by destinations. Our findings are in line with Gali et al. (2016) who explained that slogans are moving from descriptive strategies (using words related to the features of the destination) to one that appeals to the emotions (very few words). Such approaches arguably get beyond the exploitative normative approaches (neutrality) and into more explorative approaches (innovative marketing strategies through emotive marketing). Future research could consider the impact of a DMO new branding strategy on the performance of a destination.

5.7 A LOGO AS A SYMBOL

A logo is a key component of brand identity and provides instant recognition for a brand (Park et al., 2013). Because logos visually represent the essence of a brand and what the brand stands for, they have the potential to serve as a focal point of connection for customers, communicating, and reinforcing a brand's core values (Park et al., 2013). Logos are therefore a form of visual communication to customers transcending international boundaries and language barriers (Pittard et al., 2007). Yet, despite the importance of logo design in marketing and communication, little empirical research exists around this issue (Pittard et al., 2007). A Logo should be able to communicate an organization's objective to its target market, be memorable, and easily lead to an association with the corporate or brand name (Marti, 2008; Morgan et al., 2013). Logos, as visual symbols, offer an untapped opportunity (Park et al., 2013). The logo should also embody the firm's uniqueness and value and yield customer brand commitment and improve the organization's performance, while contributing to favorable interaction between the firm and its customers (Park et al., 2013; Wang et al., 2012). In a tourism context, Hem & Iversen (2004) argue that destination logos should match the destinations they represent, "their identity, uniqueness, and value, as well as essence and objective" (Marti, 2008; Morgan et al., 2013). These traits are all the more important in tourism because logos can influence tourists' destination choice (Gallarza et al., 2002).

The effectiveness of a visual logo likely depends on multiple design properties of the logo, such as color, shapes, and sizes (Park et al., 2013). Communication, marketing, and retail sectors acknowledge the role of color in attracting attention, inducing a mood, or conveying complex information and use color to achieve specific goals. The color marketing group (CMG) suggests that color increases brand recognition and accelerates learning. The role of color in logos essentially relates to its symbolic function, which has remarkable power. An accumulation of archaeological evidence over the last 10 years indicates the use of colored pigments (e.g., ochre) in pre-historical times to convey meaning (Labrecque et al., 2013).

A tourism destination usually comprises a variety of natural and cultural resources upon which a myriad of stakeholders works to co-create the final tourism offer. As a result, because of the involvement of multiple hands in creating the tourism product, determining the appropriate colors that best represent the destination is often an area of high debate among stakeholders. The choice of the color palette that best captures the essence and narrative of the destination is also an important pre-requisite (Morgan et al., 2013). Marketers have long used color to catch consumers' attention and to communicate a brand meaning, and they continually review color meanings and associations to adapt to the market in order to gain a competitive edge. Tourism marketing is a sector that has not yet optimized the use of color in its communication's strategy; DMO logos are suitable examples to illustrate this viewpoint. The implication for DMOs is that they face the challenging task of selecting the appropriate color that not only best represents the destination, but also generates a strong image in the mind of visitors. As a result, DMOs should be aware of the perceived importance of colors and their interpretation when designing a logo. Firms should conduct research regarding choice of colors before launching a product or service. Following this perspective, the performance of color in a logo (Figure 5.1), if well calculated can affect the destination image, perceived value of destination, loyalty, attractiveness, and hence competitiveness of a destination. Tourism marketing, like other marketing areas, could take advantage of the research and development in color psychology to enhance communication and effect. The framework in Figure 5.1 builds on propositions that claim that capturing the essence of the destination is very important for any visual element (Morgan et al., 2013). Thus, a translation system of parameters into colors can determine the palette of colors of the logo.

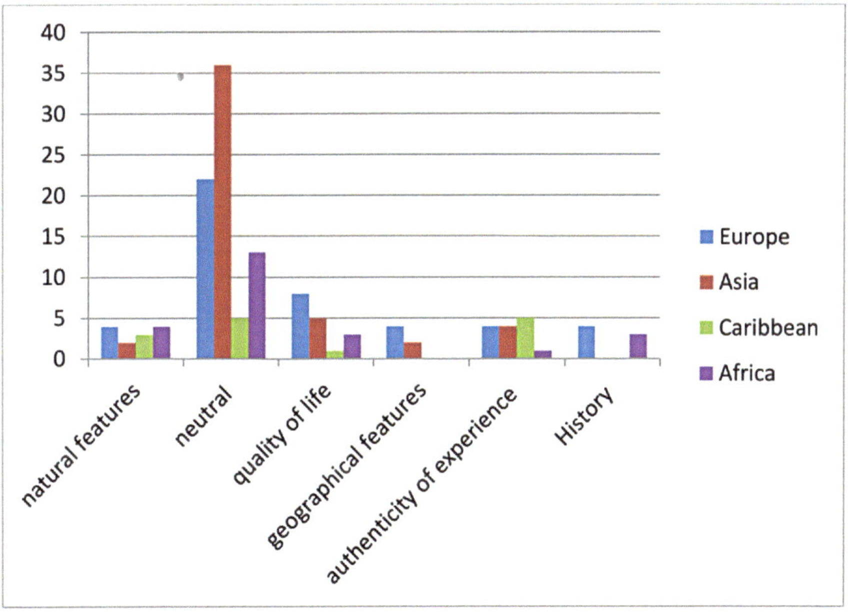

FIGURE 5.1 Performance of color in logo.

Source: Séraphin, Ambaye.

The outer circles of "*Destination*" and "*Marketing*" show all the key criteria that firms should take into account when deciding the colors of an effective logo. The framework also highlights the complexity of designing a good logo because the logo needs to simultaneously reflect the essence of the destination and break with the previous image of the destination. Applying the above model to Haiti's DMO logo, the colors in the first logo are more effective in terms of conveying the essence of the destination to the tourists, because the colors used are quite meaningful and representative of the history, tradition, identity, and context of Haiti. The new logo of the Haitian DMO, as Muller et al. (2013) argue, contributes by changing the image of the destination into something more modern. From this point of view, the logo achieves the objectives and values of the DMO. However, the logo does not give the full narrative of the destination. In terms of brand recognition, uniqueness, and memorable traits, colors of the Haitian DMO logo are not that different from other Caribbean DMO logos. Caribbean DMO logos commonly use red and yellow. The colors (and design) of the Haitian DMO logo are not very different from the logo of the Dominican Republic (DR) DMO logo. Therefore, the Haitian DMO may have used a benchmarking approach when adopting their current logo. This situation echoes Séraphin's research (2011, p. 43), which argues that "whatever direction Haiti and the DR choose to go to in order to achieve a sustainable tourism industry that is economically, geographically, and socially viable, a partnership between both countries seems to be unavoidable as there is a high risk of duplication."

5.8 'NEW ERA' DESTINATION BRANDING

COVID-19 has transformed understanding of both 'Destinations' and 'Marketing' as conceptualized in Figure 5.1. The following section underlines the finding so far that a neutral strategy is commonly adopted, in appealing to emotions, rather than use of words relating to scientific, cultural, or built heritage. Destinations across the globe are presented with the common challenge of ensuring safety and cleanliness protocols in response to the COVID-19 pandemic, and communicating them in an accessible, timely, and accurate way in order to inform consumer decision-making (World Travel and Tourism Council, 2020).

Destination image can be understood through cognitive (knowledge and beliefs), affective (feelings and attitudes) and conative (desired future situations) (Pike & Ryan, 2004; Tan & Wu, 2016; Afshardoost & Eshaghi, 2020). Differing countries' responses to the pandemic have affected destination image through media coverage (Gössling et al., 2021) which influence what consumers know and/or believe about a country, combined with the influence on feelings and attitudes through tourists' perceptions of safety and risk (as shown for example by Casali et al., 2020). Mostafa illustrates that health risk influences perception, image, and travel intention, and perceptions might be associated with past experiences, which might be especially important if perceptions have been shaped by misleading or negative information. Pre-pandemic, negative perceptions in some destinations were linked to over-tourism and resultant resident discontent and conflict (Séraphin et al., 2021), and some early media reporting attributed the spread of the pandemic to travel and tourism, which might further increase tensions, and in turn consumers cognitive and affective evaluation of destinations. Sun Wung et al. (2021) suggest that in order to overcome any potential negative perceptions, 'impression management' needs to compensate for this through developing positive views towards another aspect of the destination.

National Tourist Board responses have needed to operationally respond to health risks by demonstrating that countries are safe to visit, but also to convey that the tourists' experiences in destination will be enjoyable (rather than restrictive). Further understanding is required of consumer attitudes to biosecurity measures (such as wearing marks, testing, social distancing, isolation) in order that marketing communications can be targeted to build confidence (Ja Kim et al., 2022). Paradoxically, destination brands may for example need to convey both a sense of adventure and a sense of safety. This presents a significant opportunity for pioneering approaches. The sense of destinations appealing to tourists' emotions, while demonstrating a responsible approach to destination management is evident in Visit Portugal's '#ItsTimeToStop' campaign (website) which ran during the lockdown, and now focuses on '#ItsTimeToBe. Visit Maldives' campaign '#VisitMaldivesLater – But Keep Dreaming Now' illustrated the construction of emotional connections/bonds in communications. As the world opened up, Destination Management Organizations such as Dubai Tourism responded with the innovative '#ReadyWhenYouAre' campaign, heralding a new era of responsibility for destination marketeers in demonstrating destination hygienic safety. These factors are important in shaping affective attitudes, and also developing a sense of trust and safety to overcome any negative media reporting. For tourists without previous destination familiarity (cognitive dimensions), then such aspects of health and trust can be key indicators of behavioral intentions (Rasoolimanesh et al., 2021). In a new era of travel and marketing, destination branding for strategic decision making is perhaps more important than ever before.

5.9 DISCUSSION AND CONCLUSION

As an exploratory study, this research argues that proper choice of color is critical in developing destination logos because color is one of the main vehicles that communicate image, cut through clutter to gain attention, and speed up recognition of the destination (Hem & Iversen, 2004). As a result, tourism marketing managers must develop a better understanding of the effects of color in order to make informed decisions in the development of strong destination logos. Tourists often perceive destination image through its logo, which has been an important contributor in determining destinations' representativeness, visual brand awareness, and marketing effectiveness (Morgan et al., 2013).

Figure 5.1 provides a method to determine the right palette of colors for a logo. This process is the "nutshell approach," the nut being the core elements of the destination (context, identity, and history) and the shell the visual aspects (esthetic, the memorable aspects, etc.). The current research about the Haitian DMO goes far beyond a research study on the marketing of a postcolonial, post-conflict, and post-disaster destination, or about the role of color in marketing. This study also sheds some light on the role of the diaspora in the tourism development of the destination as a purveyor of knowledge, innovation, entrepreneurship, and development. Séraphin (2015) also explains that diaspora tourism presents more benefits for Haiti than non-diaspora tourism.

The results show all the importance of innovation in the tourism industry with competition as the main driver (Edgell & Swanson, 2013). In the same line of thought, competitiveness influences the success of a destination (Enright & Newton, 2005). The market conditions and business environment in which tourism operates are constantly changing, requiring businesses to adapt, to develop strategies and to be innovative to retain their position (Page, 2012). With globalization, the tourists have more choices, therefore increasing the competition between destinations (Holden, 2013). Performance becomes a key feature of competitiveness (Holden, 2013).

The development of new products and services such as adult only hotel (AoH) is also an approach for destination to brand themselves and standout from the crowd. The 'Adult Only' Hotel (AoH) is an emerging type of establishment based on the premise that adult indulging in wellness activities need to be freed from domestic responsibilities in order to effectively benefit from wellness programs (IHCS, 2021). AoHs and wellness tourism have similar characteristics due to their connection with spa and health (IHCS, 2021). Additionally, they are based on niche products, and both are being the opposite of mass tourism (Park, 2014). Whilst a direct link may be established between AoH and wellness tourism product, it would seem that its lucrativeness and potential have still not been explored in the context of destination competitiveness. Both wellness tourism and AoH appeared at the beginning of this century and both products are being used in diversification, product differentiation and competition (IHCS, 2021). Additionally, adult only' hotels have a balanced comparative and competitive advantages. This type of hotels assimilated with Special Interest Tourism (including of course wellness tourism) could also be a solution to overtourism, one of the most recent issues faced by the tourism industry (Dodds & Butler, 2019), as this type of hotels focus on high end customers (IHCS, 2021), as opposed to mass tourism, which is enabled among other things by affordability of tourism products and services (Dodds & Butler, 2019). By contributing to reduce the number of visitors, this type of hotels would enable destinations to yield the same income from visitors, as special interest forms of tourism, and tailor-made products and services are known to help destinations to yield a maximum of income from visitors, as they are meeting their specific needs (Favre, 2017). The development of this type of hotels could be part of a degrowth strategy in tourism which is all about a better focus on experience of locals and visitors (quality) as opposed to quantity (Dodds & Butler, 2019; Milano, Novelli, & Cheer, 2019).

KEYWORDS

- **destinations**
- **global consumption**
- **heritage**
- **social cohesion**
- **unity**

REFERENCES

Afshardoost, M., & Eshaghi, M. S., (2020). Destination image and tourist behavioral intentions: A meta-analysis. *Tourism Management, 81*, 104154.

Ashley, C., De Brine, P., Lehr, A., & Wilde, H., (2007). The role of the tourism sector in expanding economic opportunity. *Corporate Social Responsibility Initiative Report No. 23*. Kennedy School of Government, Harvard University: Cambridge, MA, USA.

Avraham, E., (2016). Destination marketing and image repair during tourism crises: The case of Egypt. *Journal of Hospitality and Tourism Management, 28*, 41–48.

Bucheli, M., (2008). Multinational corporations, totalitarian regimes and economic nationalism: United Fruit Company in Central America, 1899–1975. *Business History, 50*(4), 433–454.

Casali, G. L., Liu, Y., Presenza, A., & Moyle, C. L., (2020). How does familiarity shape destination image and loyalty for visitors and residents? *Journal of Vacation Marketing*.

Cashmore, E., (2006). *The Black Culture Industry*. London: Routledge.

Dodds, R., & Butler, R. W., (2019). *Overtourism: Issues, Realities and Solutions*. Berlin: De Gruyter.

Dubesset, E., (2013). Thinking differently about Caribbean regional identity. *Etude Caribeennes,* 21. doi: 10.4000/etudescaribeennes.5739.

Favre, C., (2017). The Small2Mighty tourism academy: Growing business to grow women as a transformative strategy for emerging destinations. *Worldwide Hospitality and Tourism Themes, 9*(5), 555–563.

Gartner, W. C., (2009). Deconstructing brand equity. In: Cai, L. A., Gartner, W. C., & Munar, A. M., (eds.), *Tourism Branding: Communities in Action* (pp. 52–64). United Kingdom: Emerald Group.

Gilmore, F., (2002). A country—Can it be repositioned? Spain—The success story of country branding. *The Journal of Brand Management, 9*(4), 281–293.

Gopalan, S., (2013). *Mapping India-EU Tourism Flows; CARIM–India Research Report No. 2013/15.* Robert Schuman Centre for Advanced Studies, European University Institute: San Domenico di Fiesole (FI), Italy.

Gössling, S., Scott, D., & Hall, C. M., (2021). Pandemics, tourism and global change: A rapid assessment of COVID-19. *Journal of Sustainable Tourism, 29*, 1–20.

Green, J. W., (1972). Culture and colonialism in the West Indies. *Journal of Interamerican Studies and World Affairs, 14*(4), 489–495.
International Hotel Consulting Services (IHCS), (2021). Retrieved from: https://www.ihcshotelconsulting.com/blog/adult-only-hotels/ (accessed on 20 January 2022).
Ja Kim, M., Bonn, M., & Hall, M., (2022). What influences COVID-19 biosecurity behavior for tourism? *Current Issues in Tourism, 25*(1), 21–27.
Malik, S., Chaudhry, I. S., & Sheikh, M. R., (2010). Tourism, economic growth and current account deficit in Pakistan: Evidence from co-integration and causal analysis. *Eur. J. Econ. Finance Adm. Sci., 22*, 21–31.
Mihalache, M., & Mihalache, O. R., (2016). Organizational ambidexterity and sustained performance in the tourism industry. *Annals of Tourism Research, 56*, 128–163.
Milano, C., Novelli, M., & Cheer, J., (2019). Overtourism and degrowth: A social movements perspective. *Journal of Sustainable Tourism, 27*(12), 1857–1875.
Park, C. W., Eisingerich, A. B., Pol, G., & Park, J. W., (2013). The role of brand logos in firm performance. *Journal of Business Research, 66*(2), 180–187.
Park, H. Y., (2014). *Heritage Tourism*. Abingdon: Routledge.
Park, H. Y., (2014). *Heritage Tourism*. London: Routledge.
Pike, S., & Ryan, C., (2004). Destination positioning analysis through a comparison of cognitive, affective, and conative perceptions. *Journal of Travel Research, 42*(4), 333–342.
Pike, S., (2005). Tourism destination branding complexity. *Journal of Product & Brand Management, 14*(4), 258, 259.
Rasoolimanesh, S. M., Seyfi, S., Rastegar, R., & Hall, C. M., (2021). Destination image during the COVID-19 pandemic and future travel behavior: The moderating role of past experience. *Journal of Destination Marketing and Management, 21*, 100620.
Reyes-Santos, I., (2013). On pan-Antillean politics: Ramon Emeterio Betances and Gregorio Luperon speak to the present. *Callaloo, 36*(1), 142–157.
Séraphin, H., Ambaye, M., Gowreesunkar, V., & Bonnardel, V., (2016). A marketing research tool for destination marketing organizations' logo design. *Journal of Business Research*. doi: 10.1016/j.jbusres.2016.04.074
Séraphin, H., Gladkikh, T., & Vo Thanh, T., (2021). *Overtourism: Causes, Implications and Solutions*. Springer Nature Palgrave Macmillan: Switzerland AG.
Smith, S. M., (2016). Management and organization – the 21st century global and international context. In: Stokes, P., Moore, N., Smith, S. M., Rowland, C., & Scott, P., (eds.), *Organizational Management: Approaches and Solutions* (pp. 1–26). London: Kogan Page.
Smith, S., (2017). Organizational ambidexterity: Welcome to paradox city. *Human Resources Management International Digest, 25*(1), 1–3.
Stokes, P., Moore, N., Moss, D., Mathews, M., Smith, S. M., & Yi-Peng, L., (2015). The micro-dynamics of intra-organizational and individual behavior and their role in organizational ambidexterity boundaries. *Human Resource Management, 54*(1), 63–86.
Sun, T. V. W., Tse, S., & Chan, D. C., (2021). Host-guest relations and destination image: Compensatory effects, impression management, and implications for tourism recovery. *Journal of Travel & Tourism Marketing, 38*(8), 833–844.

Tan, W. K., & Wu, C. E., (2016). An investigation of the relationships among destination familiarity, destination image and future visit intention. *Journal of Destination Marketing & Management, 5*(3), 214–226.

Thano, R., (2015). The impact of tourism on the balance of payments. *Am. J. Econ. Finance Manag., 1*, 529–536.

United Nations World Tourism Organization (UNWTO), (2016). *Tourism Highlights.* [online]. Available at: http://www.e-unwto.org/doi/pdf/10.18111/9789284418145 (accessed on 20 January 2022).

UNWTO/GTERC, *Annual Report on Tourism Trends – 2016 Edition, Executive Summary*, UNWTO, Madrid.

World Economic Forum (WEF), (2015). *The Global Competitiveness Report 2015–2016.* [online] Available at: http://reports.weforum.org/global-competitiveness-report-2015-2016/ (accessed on 20 January 2022).

World Travel and Tourism Council, (2020). *To Recovery and Beyond – The Future of Travel and Tourism in the Wake of COVID-19.* Available at: https://wttc.org/Portals/0/Documents/Reports/2020/To%20Recovery%20and%20Beyond-The%20Future%20of%20Travel%20Tourism%20in%20the%20Wake%20of%20COVID-19.pdf?ver=2021-02-25-183120-543 (accessed on 20 January 2022).

WTTC, https://wttc.org/Portals/0/Documents/Reports/2020/To%20Recovery%20and%20Beyond-The%20Future%20of%20Travel%20Tourism%20in%20the%20Wake%20of%20COVID-19.pdf?ver=2021-02-25-183120-543 (accessed on 20 January 2022).

CHAPTER 6

Global Branding Strategy and Framework: Exploring the Potential of Street Food as a Sustainable Livelihood Tourism Strategy for Developing Destinations

MANUELA PILATO,[1] HUGUES SÉRAPHIN,[2] ANCA YALLOP,[3] and CLAUDIO BELLIA[4]

[1]*Department of Responsible Management and Leadership, The University of Winchester, Road Winchester Hampshire, SO22 4NR, UK*

[2]*Oxford Brookes Business School, United Kingdom*

[3]*Department of International Business Strategy and Entrepreneurship, Auckland University of Technology, 55 Wellesley St E. Auckland, New Zealand*

[4]*Department of Agricultural, Food and Environment (Di3A) Section of Agri-Food Economics, University of Catania, Santa Sofia, Italy*

ABSTRACT

The most sustainable form of tourism activities is the one directly related to the daily life of locals. Hence the focus of this chapter is on street food. As a lifestyle, and as a tourism experience, street food is presented as a global phenomenon, not only present in major capitals, but also in smaller ones, involving a large number of consumers

Post-Disaster and Post-Conflict Tourism (An Updated 2nd Edition), Maximiliano E. Korstanje, Hugues Seraphin, and Vanessa G. B. Gowreesunkar (Eds.)

(local, regional, and internationalization). Having said that, as a tourism activity, street food remains quite vulnerable to external forces such as pandemics. Tourism and street food businesses were severely affected by the COVID-19 crisis due to closures and strict health regulations imposed by national authorities, worldwide. Tourism businesses have applied a wide range of responses, ranging from compliance with health measures to innovation in their services, adoption of new technologies, and the search for new customer segments to regain their customers.

6.1 INTRODUCTION

The years 2020 and 2021 will go down in history as the years of the COVID-19 pandemic. A phenomenon that not only created an unprecedented, challenging, and complicated health situation, but also significantly changed the lives of billions of people. A change that, of course, has also arrived in the tourism and agri-food sectors.

Due to the COVID-19 pandemic that spread from 2020 all countries around the world have faced and are still experiencing an unprecedented crisis.

In most countries around the world, during the periods of greatest virus spread, people remained segregated, locked in their homes, physically isolated and, for the first time, in contact with the world primarily through digital media.

Most of the inhabitants of high-density urban centers found themselves in a particularly sacrificed condition due to the confinement within the city perimeter, often characterized by the insufficiency or non-usability of green areas and modest environmental quality.

The economic impact of the pandemic has affected all the countries of the world, generating, however, extremely asymmetrical effects linked, above all, to the socio-economic disparities and the pre-conditions of the context of the various territories (Auriemma & Iannaccone, 2020).

Before the pandemic, it is recalled that the fastest growing trend for international travel has been for travel to Less Developed Countries (Holden, 2013). Taking the example of Haiti, Séraphin, Ambaye, Gowreesunkar, & Bonnardel (2016) explained that tourism is quite central in the strategy for the economic development of LDCs despite the fact these destinations struggle to establish themselves as tourist destinations. There has been a good deal of discussion as to whether tourism is a godsend or an evil for these destinations. Crick (1989) points out that an industry as complex as tourism, which involves individual, local, national, and international levels in addition to economic, social, and cultural factors cannot be consistently 'good' or 'bad' for a third world country. The impact of tourism on the contemporary world is profound. Apart from pandemic, war, and insecurity, it accounts for the largest movement of human populations. Many third world countries have chosen the tourism industry as a central development strategy, strongly encouraged in the 1960s by groups including the Organization for Economic Cooperation and Development, World Bank and the United Nations. Since the 1980s there has been a growing recognition that tourism requires more equality among all participants which has led to alternative forms of tourism where less foreign capital and more local people, food, and architecture are engaged (Crick, 1989). So far, the poor performance of developing destinations is due to: loss of international appeal; poor ratio price/quality of accommodation, excursions, restaurants; poor customer service; shopping is limited and expensive; non-existent nightlife; limited activities for tourists; port and airport taxes are dissuasive; lack of motivation of the people working in the industry; foreign airlines have abandoned the destinations; appalling social climate; increase of intraregional competition (Gay, 2012). The tourism sector is especially vulnerable to exogenous factors like COVID-19 pandemic, political instability, economic crisis, natural disasters and the outbreak of diseases (Séraphin, 2014d; Hai & Chick, 2011; Ritchie, Dorrell, Miller, & Miller, 2004). Those factors can cause destinations to decline and sometimes even totally disappear from the tourism map (Seddighi, Nuttall, & Theocharus, 2001). In this respect tourism is a problematic industry (Getz, 2008).

Figure 6.1 provides an overview of International Tourist arrivals decrease caused by COVID-19 on a global, regional, and sub-regional levels in 2021.

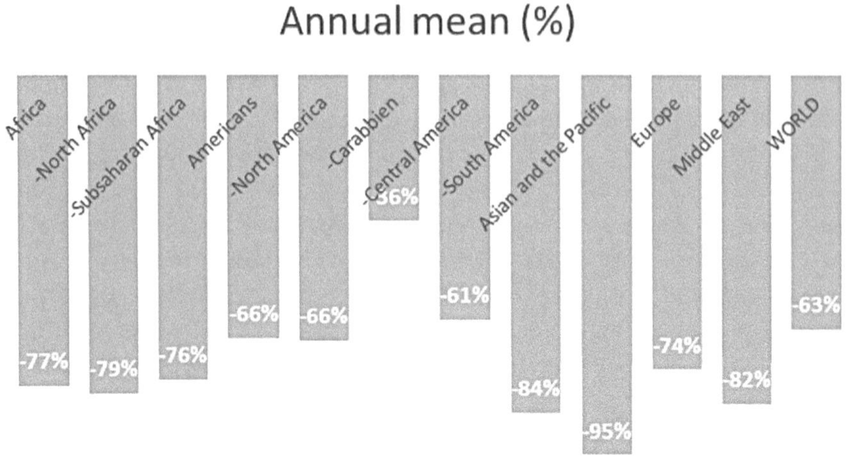

FIGURE 6.1 Assessment of the economic impact of the COVID-19 crisis on world tourism.

Note: Elaborated from UNWTO Database (2021).

In 2021, The Word Travel and Tourism Council (WTTC) conducted research to assess the impact of COVID-19 on the travel and tourism industry. Estimates show that tourism will lose $2.7 trillion in 2021, and that millions of jobs are compromised by the crisis. According to the report, the Asia-Pacific region will be the hardest hit with 63.4 million job losses, while for Europe, the research suggests a loss of tourism GDP of €633 billion.

This chapter is going to focus on an activity involving local people and local food, namely street food. This activity could be the alternative form of tourism that Crick (1989) was talking about, different from stage tourism described by Michel (2009) and alongside the authentic form of tourism in Haiti described by Thomson (2014); and Wagner (2015) where the people, the culture and natural assets are central. This authenticity is what made Haiti the most popular destination in the Caribbean between the 1940s–1960s (Theodat, 2004). The following sentences summarize the situation and key issue related with authenticity: 'Authenticity is regarded as the most important criterion for the development of heritage tourism' (Xie & Wall, 2003 in Park, 2014, p. 62) 'What tourists experience as authentic often turns out to be staged authenticity which is carefully manufactured and promoted by the local and tourism industry' (...) 'Staged authenticity in tourism seems to discourage modern tourists to search for authentic experiences' (…) 'The search for authenticity often becomes meaningless due to strategically contrived and constructed nature of tourism settings and experiences' (Park, 2014, p. 60). In such a context, Cros & McKercher (2015) explain that it is important to find a fine line between selling some aspect of a culture and keeping this culture authentic. There is a call for the tourism industry to be more ethical in the way it does business (Lovelock & Lovelock, 2013). Cros & McKercher (2015) also emphasized on the fact that the intangible heritage of a destination is a strong indicator of its authenticity.

Food (taste, ingredient, accompaniment, preparation, form of serving the dish and the way of consuming the dish), was an important motivator that push people to travel the world (Kowalczyk, 2014). In destinations like Africa, Asia, and Latin America, where it is a common way of eating for a substantial part of the population, street-food that refers to ready-to-eat-food sold in a street or a public place (Kowalczyk, 2014) could be considered as an authentic way of living and an authentic activity visitors could be involved in. Beyond being an authentic activity, street food is also a livelihood for locals. Street food was consumed each day by an estimated 2.5 billion people worldwide. In Latin America, street food purchases account for up to 30% of urban household spending. In Bangkok, 20,000 street food vendors provide city residents with an estimated 40% of their overall food intake. In Thailand, a huge percentage of the population depends on this entrepreneurial activity for survival. In Ghana, it is estimated that the sector employs over 60,000 people and has an annual turnover of US $100 million, with annual profits in the order of US $24 million (Fellow & Hilmi, 2011). The greater the poverty, the more necessity entrepreneurship there is, thus resulting in high rates of entrepreneurial activity (Reynolds, 2001). Their strategies are individual and short term. Moreover, their businesses are not operating with a strategy but as a need to survival (Van der Sterren, 2008).

From the point of view of metropolitan territories, the health crisis has affected all categories of regions, negatively and significantly affecting both the concentration of wealth in urban areas, as well as the use and economic-social valorization of the cultural and tourist heritage, to the point of accentuating those critical environmental and quality of life issues within the large places of population concentration. Moreover, particularly in such urban contexts, due to the problems linked to the virus congestion, even the consumption of food has significantly changed. In fact, because of the persistence of lockdowns and the fear of the spread of contagions, consumers reduced the purchase of street food. People have increased food purchases at supermarkets, avoiding traditional neighborhood markets where fresh food is generally sold and there is a greater chance of infection. In addition, the request to receive groceries at home has increased considerably, being unable to go out during lockdown periods, but it has been observed that this habit has largely remained even after the reopening.

We believe all that contributes to keep the authenticity of street food in developing destinations. Our objective in this research chapter is to find out how street food can retain its authenticity in developing countries and be at the same time a profitable activity for the locals. This conceptual research chapter is therefore going to contribute to the literature in the field of tourism by identifying the fine line between selling some aspects of a culture and keeping it authentic. In that instance, this chapter could be considered a prolongation of Cros & McKercher (2015) research, however their study does not provide a way to identify this fine line. This research chapter could also be considered as a prolongation of Séraphin, Ambaye, Gowreesunkar, & Bonnardel (2016) who show the complexities and difficulties of choosing the right colors for the logo of a tourist destination when this organization have to change its image while maintaining its essence. Once again, the authenticity aspect and the commercial aspect were opposed. This research also fails to identify a fine line between both. The answer to the objective of this chapter is going to contribute to the literature in (tourism) marketing. Our hypothesis is that this fine line when identified will need to be advertised using a subtle marketing strategy as aggressive marketing promotional strategy for developing destinations often fail its objective (Alvarez & Campo, 2014; Seddighi, Nuttall, & Theocharus, 2001). The research objective of this chapter can be summarized as in Figure 6.2.

This conceptual method uses a multi-disciplinary approach combining literature on marketing research on tourist destinations; sustainability in tourism; sustainable food; street food; and entrepreneurship in tourism. This method may potentially help in deciding how to find a fine balance between keeping street food an authentic product for tourists while being profitable for locals.

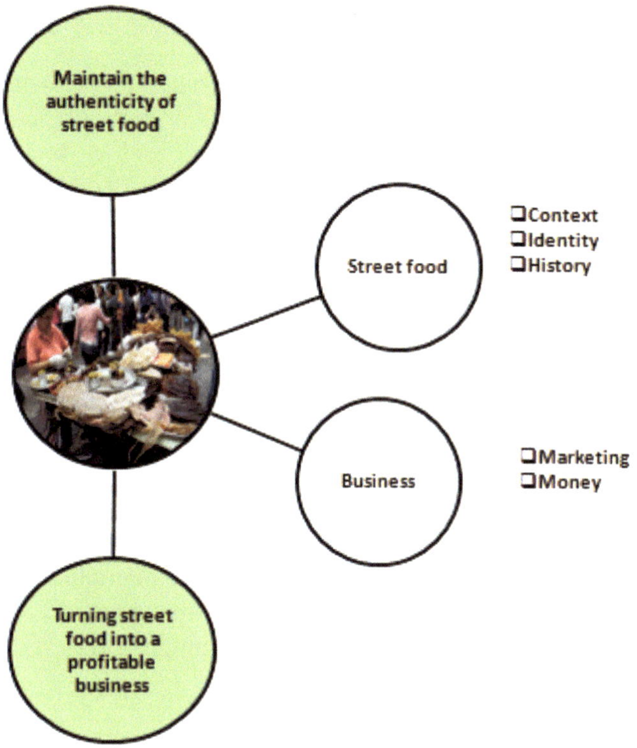

FIGURE 6.2 Research objectives.

Source: Pilato, Séraphin, & Yallop (2016).

The study has four main sections. The contextual framework sets the backdrop for the research. In the present case, the context is developing destinations, which happens to be as well destinations where street food is part of the lifestyle. The conceptual framework presents the literature published in the area of the research (Quinlan, 2011). The discussion section presents the results, such as what should be taken into account when designating the fine line between keeping a product authentic and promoting/selling the product. Finally, the conclusion focuses upon the current status of street food, providing suggestions for future research.

6.2 DEVELOPING DESTINATIONS

Emerging markets are those lower-income but rapid-growth countries that are using economic liberalization as their primary engine of growth (Hoskinsson, Eden, Lau, & Wright, 2000). The low socio-economic development of these countries was often due to political instability, great deal of conflicts during a short period, no sense of national identity among the people, economic crisis, natural disasters, and the outbreak of diseases (Gould, 2011; Ritchie, Dorrell, Miller, & Miller, 2004). The emerging countries are most of the time, postcolonial, post-conflict, or post-disaster (PCCD) destinations (Bayeh, 2015; Séraphin, 2014; Gould, 2011). They also fall into two groups: first, developing countries in Asia, Latin America, Africa, the Middle East, and then, transition economies in the former Soviet Union and China (Hoskinsson et al., 2000). Séraphin, Ambaye, Gowreesunkar, & Bonnardel (2016) explained that tourism is quite often central in the strategy for the economic development of these countries. Indeed, the fastest growing trend for international travel has been for travel to less developed countries and emerging destinations (Holden, 2013; Korstanje, 2009; Alvarez & Campo, 2014; Tarlow, 2014).

6.3 SUSTAINABLE TOURISM

The tourism industry has been highlighted as an industry that can positively contribute to the economic and social development of a destination (Buckley, 2012). However, this industry can also negatively impact on a destination if poorly managed (Mazanec, Wober, & Zins, 2007 cited in Iniesta-Bonnillo, Sanchez-Fernandez, & Jimenez-Castillo, 2016). Among the noticed negative impacts of tourism are the emerging challenges for the planet; the over-exploitation of resources like water, minerals, oil, etc.; over-population (Sloan, Legrand, & Chen, 2013). As a result, a destination is considered to be sustainable if the tourism industry does not impact negatively on the environment, on human-environment interactions and local communities; but equally, the industry needs to contribute to cultural exchange between locals and visitors and meeting the economic needs of the populations (Mbaiwa, 2005 in Iniesta-Bonnillo et al., 2016). To this list could be added: the long-term capacity of the industry to remain 'clean' from an environment point of view with the support of technological systems; a fair impact on all members of the population in the present and in the future; the development of policies at local, national, and international level (Sharpley, 2000). The main concern of the development of tourism is for developing nations and Indigenous people as the impacts of the industry can be either very positive or very negative (Buckley, 2012). Hence initiatives like sustainable tourism-eliminating poverty program (ST-EP) initiated by the United Nations World Tourism Organization (UNWTO) taken to use tourism to reduce poverty and increase net benefits for poor people as part of their 'Pro-Poor Tourism' policy (Holden, 2013). Destinations now consider sustainability as a competitive advantage to attract visitors, hence the growth of ecotourism, the fastest-growing tourism sector globally (Iniesta-Bonnillo et al., 2016; Orams, 1995 cited in Kazeminia, Hultman, & Mostaghel, 2016). Consumers are ready to pay more to spend their holidays in destinations considered as sustainable (Kazeminia et al., 2016). That said, much of the tourism industry so far has failed to be sustainable (Higgins-Desbiolles, 2010 cited in Iniesta-Bonnillo et al., 2016), hence the reason why Sharpley (2000, p. 1) claimed that 'sustainable development cannot be transposed onto the specific context of tourism.' Around 12 years later the United Nations (2012, cited in Iniesta-Bonnillo et al., 2016, p. 1) claimed that: 'despite efforts to promote more sustainable tourist destinations, room for improvement exists in most countries.' This seems to suggest that sustainability in the tourism industry is very hard to achieve, despite a strong will of academics since the late 1980s (Liu, 2003) and practitioners to develop policies and practices in the area (Sharpley, 2000). The wide variety of definitions of 'sustainable development' (more than 70) makes it even more difficult (Sharpley, 2000; Garrod & Fyall, 1998). In developing countries, it is even harder to implement sustainable tourism because of the socio-economic and political context (Tosun, 2001). As a result, Liu (2003), provided a list of issues related to sustainable tourism needed to be addressed: tourist demands in terms of sustainable products have not been taken into account; sustainability is mainly analyzed from the preservation and conservation angle; the repartition of the benefits among stakeholders is poorly researched; most research focused on the fact that tourism has negative impacts on culture and the social life of the locals; research has focused on determining a threshold to tourism growth; and finally, despite the fact ecotourism, alternative tourism, responsible tourism, soft tourism, low impact tourism and community tourism have proven not to be the way forward for sustainable tourism, writers, and practitioners are still promoting these forms of tourism as being the way forward. Holden (2013) also added that there is little data that inform us of the beneficiary impacts of tourism development on the poor. Our research can be considered as a continuity of Sharpley (2000); and Tosun (2001) as our study is based on tourism development in developing destination. This research can also be considered as a continuity of (Kazeminia et al., 2016; Iniesta-Bonnillo et al., 2016) as we are going to evaluate the potential of street food as a factor of appeal and sustainable development for developing destinations. And finally, this research can be considered as a continuity of Liu (2003) as our hypothesis is that street food corresponds to: (i) a tourist demand of authenticity; (ii) does not impact negatively on locals' culture but instead is shared with visitors; (iii) the vendors in general the poor people benefit directly from this form of tourism related activity; (iv) this related tourism activity is an add on what is promoted by the destination marketing organization (DMO), therefore is not the main reason why people are visiting the destination, therefore, won't impact on the carrying capacity of the destination; and finally (v) street food is part of the way of life of some destination, therefore cannot be classified in one of the form of tourism, hence the reason we are referring to street food as a 'related tourism activity' and not as a form of tourism. In this chapter, our main argument is that related tourism activities that are components of the daily life of the locals are the most sustainable one. Street food is used here as a case study to back up our claim. This research is also to some extent testing the model developed by Bellia, Pilato, & Séraphin (2016) in Figure 6.3.

The topic of our research and the selected case study are all the more important as despite the fact tourists want to limit their negative impacts on the destinations they are visiting (Xu & Fox, 2014), they still engage in activities which negatively impact on the destination. This is all the more the case for people who actively engage in environmental protection at home. This is what Juvan & Dolnicar (2014) called attitude-behavior gap. That said, street food appears as a good 'contestant' in terms of sustainability as, on the one hand 'community attachment and community involvement are critical factors that affect the level of support for sustainable tourism development' (Lee, 2013, p. 37); 'The sustainable development of tourism requires harmonious relationships between communities, the industry and tourists (Zhang, Inbakaran, & Jackson, 2006). On the other hand, street

FIGURE 6.3 (Street) food tourism and (developing) destinations.
Source: Pilato, Bellia, & Séraphin (2016).

food appears as a way of living, a component of the locals' life. Cros & McKercher (2015) explain that street food enables locals and visitors to participate in the same type of activities as a result, they engage in local intangible cultural experience. In the same line of thought, Holder (2013) added that facilitating the direct sales of goods and services supplied by the poor to the tourist is an effective mechanism to reduce poverty. India 2.5% of the population is represented by street food vendors; in Latin America, street food accounts for up to 30% of the expenditure of urban households and in Thailand there are an estimated 20,000 street food vendors (FAO, 2009). Food is an important motivating factor for tourists to visit a destination and occupy one third of their expenditure when on holiday (Mark, Lumbers, & Eves, 2012 cited in Lee, Packer, & Scott, 2015). It is therefore useful to explore how a livelihood could be incorporated into a destination tourism offer. Street food, therefore, livelihood will become a form of diversification. In this instance, our research can also be considered as a continuity of Tao & Wall (2009, p. 91) research as they explored 'how tourism is and might be incorporated into existing mix livelihood strategies so that it enriches rather than replaces the means by which people may be sustained.' We are analyzing the situation the other way round and providing a practical application of implementing Sustainable Tourism. At the moment, method of delivering Sustainable Tourism is not fully explored (Waligo, Clarke, & Hawkins, 2013).

6.4 SUSTAINABLE FOOD

As 'there is no international legal definition of 'sustainable food' (Sloan, Legrand, & Chen, 2013, p. 170), using Sloan et al. (2013, pp. 170–178) research, we are in this chapter, defining sustainable food in a developing tourist destination context as food that:

- Provides a viable livelihood for farmers, processors, and retailers;
- Supports rural economies and the diversity of rural culture, in particular though an emphasis on local products that keep food miles to a minimum;
- Use local, seasonally available ingredients as standard;
- Reduce processed food;
- Reduce packaging;
- Support local biodiversity and local resources;
- Serve insects as food;
- Ethical by not using endangered species, or species grown in cruel conditions.

According to the UN World Tourism Organization (2012) tourist choices are increasingly influenced by sustainability considerations. Local foods and drinks are an asset to sustainable tourism – they enable host communities to capitalize on visitor's desire for authenticity (Sims, 2009).

Street food is noted to be low in cost and present an attractive alternative to home-cooked food. Street food vending is usually practiced as a family business and in the majority of cases; it is a source of employment for the household members (FAO, 2014). According to the Codex International Food Guidelines (1999) street food are ready-to-eat foods prepared and/or sold by vendors and hawkers, especially in streets and other similar public places. While according to WHO (2006), street food is ready-to-eat foods or beverages prepared and/or sold in the street and other public places for immediate consumption or at a later time without further processing or preparation.

Other public places include schools, markets, and motor parks (Muleta & Ashenafi, 2001). According to the FAO definition (2009) given "street food is made from food and drink prepared for consumption, prepared, and sold on the street or in other public places such as markets and/or exhibitions, often sold on counters or through temporary vans and carts street vendors."

A street food vendor is broadly defined as a person who offers foods for sale to the public without a permanent built up structure but with a temporary static structure or mobile stall-head load/wheel-barrow/truck (Janie & Marie, 2010). Street vended foods provide a source of inexpensive, convenient, and often nutritious food for urban and rural poor; a source of attractive and varied food for tourists and the economically advantaged; a major source of income for a vast number of persons, particularly women; and a chance for self-employment and the opportunity to develop business skills with low capital investment (Codex International Food Guidelines, 1999). Street food, then, can be considered not only a phenomenon of fashion since the act of eating on the street is part of the mentality of the past generations, due in search of food less expensive and at the same time fast (Bellia et al., 2016).

In addition, as noted, goal 8, one of the United Nations global goal for sustainable development, promote sustainable tourism, in particular Target 8.9 'devise and implement policies to promote sustainable tourism that creates jobs and promotes local culture and products' (UN, 2015).

According to Marras (2014 cited in Global Forum on Food Security and Nutrition), street food account for a significant proportion of daily urban food consumption for millions of low- and middle-income consumers in urban areas, representing the least expensive and most accessible means of obtaining a nutritionally balanced meal outside the home. As mentioned above, the preparation and sale of street food provides a regular source of income for millions of men and women with limited financial, social, and cultural capital, since the start-up investment and the overheads are relatively low, and cooking requires little or no formal training. A study by Sautier, Vermeulen, Fok, & Biénabe (2006) reveals that in Garoua, a city in Cameron, of the 781 street food enterprises present, all were set up and operated without any financial intervention either from the public sector or from donors.

Moreover, 'reports from a number of countries have shown that income from street food vending can be very good' (Fellows & Hilmi, 2011, p. 16). Street food at global level provide a wide range of commodities and nutrients, helping people to meet their nutritional needs: indeed, easy accessibility, variety in taste and choice and low cost make street foods an affordable option even for tourists (Marras, 2014 cited in Global Forum on Food Security and Nutrition).

However, research shows that nutritional and sustainable diets cannot avoid the food safety aspect of foods, it is interesting to note that research also shows that street foods play an important socio-economic role, in terms of employment and income for those involved in their production and/or selling, empowerment of the local economy and the preservation of the local food cultures (Proietti cited in Global Forum on Food Security & Nutrition, 2014; Bellia et al., 2022). Within this framework, street food vendors in developing countries can provide fresh, nutritious, less expensive ingredients which enable host communities to capitalize on visitor's desire for authenticity (Sims, 2009).

Although in most cities in developing countries, an important part of urban agricultural production is for self-consumption, the importance of the market-oriented street food tourism, both in volume and economic value, should not be underestimated. For example, local food production can be an important source of supply of fresh vegetables for street food preparation. Since locally produced food requires less transportation and refrigeration, it can supply nearby markets with fresher and more nutritious products at competitive prices. Food growing in cities, thus, can and does help improve the quality of people's diets by providing a greater choice of fresh fruits and vegetables at better prices, particularly to people in the low-income bracket. Common street foods, in most countries, are based on animal-source ingredients often derived from animals kept in cities, there where the commercial peri-urban production of livestock is an extremely fast-growing sector, representing 34% of total meat production and nearly 70% of egg production worldwide (Marras, 2014 cited in Global Forum on Food Security and Nutrition). Indeed, fresh salad preparations have increasingly become an integral component of street food sold in cities like Accra, Ghana, and Santiago, Chile.

However, the food sold on the street may also represent a risk. As a consequence, research shows greater awareness and preventive measures need to be implemented for coping with risk factors in a systematic and effective way. Food safety involves the protection of supply from microbial, chemical, and physical hazards that may occur during all stages of food production, including growing, harvesting, processing, transporting, retailing, distributing, storing, preparing, and consuming to prevent food borne diseases (WHO, 2006). For example, the Government of Bangladesh has recognized the role of street foods in urban food security and it has been outlined as a strategic area of intervention in the National Food Policy of Action (2008–2015) under strategy 3.5, entitled as "*Safe, Quality Food Supply: Institutionalization of Safe and Hygienic Street Food Vending as Medium and Long Term Actions*" (FAO, 2014). Precise numbers of street food vendors in Bangladesh are not available, but estimates show that the capital Dhaka alone (population: 14 million) may have more than 2 million street vendors (FAO, 2014).

6.5 ENTREPRENEURSHIP TOURISM

Michel (2000) claims tourism can fully benefit a destination, only if the locals are fully involved in the planning and development of the industry. Based on the literature, entrepreneurship in tourism seems to be a serious contender to the involvement of the locals in the tourism sector and in some cases to help them to improve their standard of living. In fact, according to Manyara & Jones (2005), micro- and small-scale enterprises can increase the participation of the poor in the tourism industry. Rogerson (2003) based on his research in South Africa backs up Manyara & Jones (2005) as he argues that it is only through small enterprises that rural people can participate in tourism. Literature not only shows entrepreneurship is a factor for economic development (Schumpeter, 1934), but it can also have a multiplier effect in some cases on local economies. When formal tourism enterprises are owned by locals, there is a high likelihood of purchase of local supplies, meaning other sectors of the local economy are going to benefit from this activity Shah (2000). Wanhill (2000) supports this point of view as he contends that for tourism to be beneficial in terms of income generations, economic growth, poverty reduction, and improving rural livelihoods, it has to be linked to local economies such as agriculture and MSEs. Butler (1980) views entrepreneurial development by local residents as occurring spontaneously in response to growth in tourist demand. Theodat (2004) shows in his research on Haiti & Thomson (2004), in his novel: '*Bonjour blanc, a journey through Haiti*' that this entrepreneurial development mainly happened in the hospitality sector. We can complete Butler's theory (1980) by saying that, when there is a growth in tourist demand, entrepreneurial development occurs in specific sectors of the tourism sector and those developments are conducted by local residents with an established financial situation. However, in some cases, those residents who could have encouraged the economic development of the country do not want or cannot do it. Din (1992) noted that the receiving community may not possess the cultural and economic capacity to appreciate the opportunities. Kristen & Rogerson (2002) noted the same situation in their research on tourism MSEs in South Africa where there is a limited involvement of local black entrepreneurs due to their limited resources and difficulties to secure credit from financial institutions and also because of limited and inadequate training and education as most of the entrepreneurs only have primary level of education. The contemporary tourism industry is dominated by MSEs and particularly micro-businesses. Quite often they are owned and run by people inexperienced in business, contributing therefore to an 'under-management' of the tourism industry (Buhalis & Cooper, 1998; Cooper & Hall, 2008).

6.6 DISCUSSION

The aim of this chapter is to share perspectives on actual and potential links between street food as a sustainable livelihood tourism strategy for developing destinations, analyzing if and in which way such links may represent the basis for possible strategies to enhance street food tourism in developing areas, as showed in Table 6.1.

However, as previously discussed, street food is to be associated with the topic of food safety (Bellia et al., 2022). In some destinations like Haiti it is considered as a major issue as the consumers of street food and the street food vendors have a poor knowledge of the issues related to lack of food safety (Samapundo, Climat, Xhaferi, & Devlieghere, 2014). That said, the context of some destinations and the social situation of the vendors do not allow them sometimes to meet the requirements of a

TABLE 6.1 Multi-Criteria Analysis of the Sustainability of Street Food as a Tourism-Related Activity

		Bellia et al. (2016) Model – P11-17					
		Low-Cost Investments	**Quick Return on Investments**	**(Street) Food Tourism**	**Local (Food) Habit**		
Sustainable Tourism	No negative impact on the environment/human interaction/local communities	–	–	X	X	Daily life of the locals	Street Food
	Contribute to cultural exchange local visitors	–	–	X	X	Authentic	
	Meeting economic needs of the local	X	X	–	–	Family business/ major source of employment	
	Long term ability to remain clean environmentally	–	–	–	X	Animal-source ingredient	
	Development of policies	X	X	–	–	Development of business skills	
		Ethical and support rural economies	Support local biodiversity/local resources	Reduce packaging and food processed	Support local culture		
		Sustainable Food					

Source: Pilato, Bellia, & Séraphin (2016).

suitable safety context as described by Samapundo et al., 2014). Based on the fact that Haiti has no safety legislations (Samapundo et al., 2014), it means that the check list used by Sampundo et al. (2014) is based on the legislation of another country. Does that mean that when it comes to health and safety, 'one size fits all'? Can it be argued that the application of external norms or legislations could negatively impact on the authenticity of street food?

This discussion will be an opportunity to expand future research and strengthen the network of specialists involved in street food tourism and governance worldwide. *How solutions for new way of experiencing street food?*

6.7 CONCLUSION

Tourism and street food businesses were severely affected by the COVID-19 crisis due to closures and strict health regulations imposed by national authorities, worldwide. In the tourism industry, several strategies have been adopted to gradually reopen. Not so for street food vendors. In fact, tourism businesses have applied a wide range of responses, ranging from compliance with health measures to innovation in their services, adoption of new technologies and the search for new customer segments.

Initial evidence from this research indicates that the measures adopted by tourism businesses to respond to the COVID-19 crisis have different impacts on economic (profits and revenues) and non-economic (customer loyalty) performance. The data showed an overall [in]successful performance for street food vendors, as it did not encourage people to visit more tourist destinations and go out to eat.

For everything described above, the street food sector has been strongly negatively affected from an economic point of view, not only for the lower number of local consumers but also for the absence of tourists, which represent a significant market segment for street food sellers, especially in developing countries.

Street food is presented as a global phenomenon, not only present in major capitals, but also in smaller one, involving a large number of consumers, both young and not only (Bellia et al., 2016). Street food can be found in nearly every corner of the world and have been on sale for thousands of years (Fellows & Hilmi, 2011).

This study examined the literature in the field of tourism and business ethics by identifying the fine line between selling some aspects of a culture and keeping it authentic.

The answer to the objective of this research contributed to the literature in (tourism) marketing as well as in business ethics. In particular, our study was based on tourism development in developing destination. As previously discussed, this research aimed to be considered as a continuity of Kazeminia et al. (2016); and Iniesta-Bonnillo et al. (2016) as we evaluated the potential of street food as a factor of appeal and sustainable development for developing destinations. Moreover, the study was a continuity of Liu (2003) as our hypothesis was that street food corresponds to: (i) a tourist demand of authenticity; (ii) does not impact negatively on locals' culture but instead is shared with visitors; (iii) the vendors in general the poor people benefit directly from this form of tourism-related activity; (iv) this related tourism activity is an add on what is promoted by the DMO and the United Nations global goal for sustainable development, Goal 8 which promote sustainable tourism. Therefore, is not the main reason why people are visiting the destination, therefore, will not impact on the carrying capacity of the destination; and (v) finally street food is part of the way of life of some destination, therefore cannot be classified in one of the form of tourism, hence the reason we were referring to street food as a 'related tourism activity' and not as a form of tourism. It was also particularly noted that related tourism activities which are components of the daily life of the locals are the most sustainable one. For these reasons, street food was used in this research as a case study to back up our claim as well testing the model developed by Bellia, Pilato, & Séraphin (2016) which showed that tourism can positively impact on challenges faced by a destination and maximize their opportunities. The study recommends to expand future research and strengthen the network of specialists involved in street food tourism and governance worldwide. In particular related to social marketing; social mobilization; development support communication (DSC); socioeconomic impact; legislative framework; hygienic and nutritional improvement through education and training. Finally, share experiences and promote networking among local and national authorities at the regional level to spread good practices and promote a strategic coalition towards street food tourism in order to operate effectively and efficiently.

KEYWORDS

- **experience**
- **lifestyle**
- **street-food**
- **sustainability**
- **vulnerable**

REFERENCES

Alvarez, M. D., & Campo, S., (2014). The influence of political conflicts on country image and intention to visit: A study of Israel's image. *Tourism Management, 40*, 70–78.

Auriemma, V., & Iannaccone, C., (2020). COVID-19 pandemic: Socio-economic consequences of social distancing measures in Italy. *Frontiers in Sociology*, 5. https://doi.org/10.3389/fsoc.2020.575791.

Bayeh, E., (2015). The legacy of colonialism in the contemporary Africa: A cause for intrastate conflicts. *International Journal of Innovative and Applied Research, 3*(2), 2329.

Bellia, C., Bacarella, S., & Ingrassia, M., (2022). Interactions between street food and food safety topics in the scientific literature—A bibliometric analysis with science mapping. *Foods, 11*. https://doi.org/10.3390/xxxxx.

Bellia, C., Pilato, M., & Sèraphin, H., (2016). Street food and food safety: A driver for tourism? *Quality - Access to Success, 17*(S1), 20–27, ISSN: 1582-2559.

Buckley, R., (2012). Sustainable tourism: Research and reality. *Annals of Tourism Research, 39*(2), 528–546.

Buhalis, D., & Cooper, C., (1998). Competition or co-operation? Small and medium size tourism enterprises at the destination. In: Laws, E., (ed.), *Embracing and Managing Change in Tourism* (pp. 324–346). London: Routledge.

Butler, R. W., (1980). The concept of a tourist area cycle of evolution: Implications for management of resources. *Canadian Geographer, 24*(1), 5–12.

Codex International Food Guidelines, (1999). *Revised Regional Guidelines for the Design of Control Measures for Street-Vended Foods in Africa* (p. 22). CAC/GL-22R. WHO/FAO, Rome.

Cooper, C., & Hall, M., (2008). *Contemporary Tourism: An International Approach*. Elsevier: Oxford.

Crick, M., (1989). Representations of international tourism in the social sciences: Sun, sex, sights, savings, and servility. *Annual Review of Anthropology, 18*, 307–344.

Cros, H. D., & McKercher, B., (2015). *Cultural Tourism*. Abingdon: Routledge.

Din, K. H., (1992). The involvement stage in the evolution of a tourist destination. *Tourism Recreation Research, 17*(1), 10–20.

FAO, (2014). *Global Forum on Food Security and Nutrition*. Available at: http://www.fao.org/fsnforum/activities/discussions/Street-food-UPA (accessed on 20 January 2022).

FAO/WHO, (2009). *CODEX ALIMENTARIUS - Food Hygiene Basic Texts* (4th edn.). Rome: FAO/WHO.

Fellow, P., & Hilmi, M., (2011). *Selling Street and Snack Foods*. FAO diversification booklet 18.

Garrod, B., (1998). Beyond the rhetoric of sustainable tourism? *Tourism Management, 19*(3), 21.

Gay, J. C., (2012). Why is tourism doing poorly in overseas France? *Annals of Tourism Research, 39*(3), 1634–1652.

Global Forum on Food Security and Nutrition, (2014). *Street Food and Urban and Periurban Agriculture and Horticulture: Perspectives for a Strategic Coalition Towards Food Security*. Discussion No. 110.

Gould, M., (2011). Branding a post-conflict destination: Northern Island. In: Morgan, N., & Green, J. W., (1972). *Journal of Interamerican Studies and World Affairs, Military and Reform Governments in Latin America, 14*(4), 489–495.

Hai, A., & Chick, A. R., (2011). *Political Stability: Country Image for Tourism Industry in Bangladesh*. International Conference on Social Sciences, Economics and Art (ISC) in January 2011 in Malaysia.

Holden, A., (2013). *Tourism, Poverty and Development*. Abingdon: Routledge.

Hoskinsson, R. E., Eden, L., Lau, C. M., & Wright, M., (2000). Strategy in emerging economies. *Academy of Management Journal, 43*(3), 249–267.

Iniesta-Bonillo, M. A., Sanchez-Fernandez, R., & Jimenez-Castillo, D., (2016). Sustainability, value, and satisfaction: Model testing and cross-validation in tourist destinations. *Journal of Business Research, 69*(11), 5002–5007.

Janie, P., & Marie, T., (2010). *Street Food in Kolkata – A Hygienic Perspective Project in Practice 400040, Blok 3.*

Juvan, E., & Dolnicar, S., (2014). The attitude-behavior gap in sustainable tourism. *Annals of Tourism Research, 48*, 76–95.

Kazeminia, A., Hultman, M., & Mostaghel, R., (2016). *Why pay more for sustainable services? The case of ecotourism. Journal of Business Research, 69*(11). 4992–4997.

Kirsten, M., & Rogerson, C., (2002). Tourism, business linkages and small enterprise development in South Africa. *Development South Africa, 19*(1), 29–59.

Korstanje, M. E., & Tarlow, P., (2012). Being lost Tourism, risk and vulnerability in the post-'9/11'entertainment industry. *Journal of Tourism and Cultural Change, 10*(1), 22–33.

Korstanje, M., (2009). Re-visiting risk perception theory in the context of travel. *E-Review of Tourism Research, 7*(4), 68–81.

Kowalczyk, A., (2014). *From Street Food to Food Districts – Gastronomy Services and Culinary Tourism in an Urban Space*. [Online] www.turystykakulturowa.org *(accessed on 20 January 2022).*

Lee, K. H., Packer, J., & Scott, N., (2015). Travel lifestyle preferences and destination activity choices of slow food members and non-members. *Tourism Management, 46*, 1–10.

Lee, T. H., (2013). Influence analysis of community resident support for sustainable tourism development. *Tourism Management, 34*, 37–46.

Lingelbach, D., De La Vina, L., & Asel, P., (2005). *What's Distinctive About Growth-Oriented Entrepreneurship in Developing Countries?* University of San Francisco.

Liu, Z., (2003). Sustainable tourism development: A critique. *Journal of Sustainable Tourism, 11*(6), 459–475.

Manyara, G., & Jones, E., (2005). Indigenous tourism SME sector in Kenya. In: Jones, E., & Haven-Tang, C., (eds.), *Tourism SMEs, Service Quality and Destination Competitiveness* (pp. 60–72). CABI Publishers.

Michel, F., (2009). Guests and others. Tourism and alterity. *Magazine Spaces,* 171.

Muleta, D., & Ashenafi, M., (2001). Salmonella, shigella and growth potential of other food-borne pathogens in Ethiopian street vended foods. *East Afr. Med. J., 78*(11), 576–580.

Park, Y. P., (2014). *Heritage Tourism*. Abingdon: Routledge.

Quinlan, C., (2011). *Business Research Methods.* Andover: South-Western CENGAGE Learning.

Reynolds, P. D., (2001). *GEM, 2001 Executive Report.* London Business School and Babson College.

Ritchie, B. W., Dorrell, H., Miller, D., & Miller, G. A., (2004). Crisis communication and recovery for the tourism industry. *Journal of Travel and Tourism Marketing, 15*(2, 3), 199.

Rogerson, C. M., (2003). Tourism and transformation: Small-enterprise development in South Africa. *Africa Insight, 33*(1, 2), 108–115.

Samapundo, S., Climat, R., Xhaferi, R., & Devlieghere, F., (2014). Food safety knowledge, attitudes and practices of street food vendors and consumers in Port-au-Prince, Haiti. *Food Control.* doi: Org/10.1016/j.foodcont.2014.09.010.

Sautier, D., Vermeulen, H., Fok, M., & Biénabe, E., (2006). *Case Studies of Agri-Processing and Contract Agriculture in Africa.* Rimisp.

Seddighi, H. R., Nuttall, M. W., & Theocharous, A. L., (2001). Does cultural background of tourists influence the destination choice? An empirical study with special reference to political instability. *Tourism Management, 22*, 181–191.

Séraphin, H., (2014). Developing a destination image: A way forward for tourism in Haïti. In: Roselé, C. P., & Raboteur, J., (eds.), *The Challenges of Tourism Development and the Economic Intelligence of Territories: Cities on Sea and on Land.* Paris: Publibook.

Séraphin, H., (2014). *Tourism: Opening for the People of Toussaint?* Paris: Publibook.

Séraphin, H., Ambaye, M., Gowreesunkar, V., & Bonnardel, V., (2016). A marketing research tool for destination marketing organizations logo design. *Journal of Business Research.* doi: http://dx.doi.org/10.1016/j.jbusres.2016.04.074.

Shah, K., (2000). *Tourism, the Poor and Other Stakeholders: Asian Experience*. ODI Fair Trade Paper: London: ODI.

Sharpley, R., (2000). Tourism and sustainable development: Exploring the theoretical divide. *Journal of Sustainable Tourism, 8*(1), 1–19.

Shumpeter, J., (1934). *The Theory of Economic Development.* Oxford: Oxford University Press.

Sims, R., (2009). Food, place and authenticity: Local food and the sustainable tourism experience. *Journal of Sustainable Tourism, 17*(3), 321–336.

Sloan, P., Legrand, W., & Chen, J. S., (2013). *Sustainability in The Hospitality Industry. Principles of Sustainable Operations*. Abingdon: Routledge.

Tao, T. C. H., & Wall, G., (2009). Tourism as a sustainable livelihood strategy. *Tourism Management, 30*, 90–98.

Tarlow, P., (2014). *Tourism Security: Strategies for Effectively Managing Travel Risk and Safety*. Elsevier.

Théodat, J. M., (2004). The Front and Behind the Scenes: The Comparative 'Touristicity' of Haiti and the Dominican Republic. Third World Review, *45*, 178.

Thomson, I., (2014). *Bonjour Blanc: A Journey Through Haiti*. Londres: Vintage.

Tosum, C., (2001). Challenges of sustainable tourism development in the developing world: The case of Turkey. *Tourism Management, 22*, 289–303.

Van, D. S. J., (2008). Financial markets, microfinance and tourism in developing countries. *ARA (Caribbean) Journal of Tourism Research, 2*(4), 35–43.

Wagner, L., (2015). *Hold Tight, Don't Let Go*. New-York: Amulet Books.

Waligo, V. M., Clarke, J., & Hawkins, R., (2013). Implementing sustainable tourism: A multi-stakeholder involvement management framework. *Tourism Management, 36*, 342–353.

Wanhill, S., (2000). Small and medium tourism enterprises. *Annals of Tourism Research, 27*, 148–163.

Xu, F., & Fox, D., (2014). Modelling attitudes to nature, tourism and sustainable development in national parks: A survey of visitors in China and the UK. *Tourism Management, 45*, 142–158.

Zhang, J., Inbakaran, R. J., & Jackson, M. S., (2006). Understanding community attitudes toward tourism and host-guest interaction in the urban-rural border region. *Tourism Geographies, 8*(2), 182–204.

CHAPTER 7

Unlocking Changes for Sports Tourism Products Using the Blakeley-Visser Model: Application to Sport Events in Small Islands: The Case of Martinique

HUGUES SÉRAPHIN[1] and VANESSA G. B. GOWREESUNKAR[2]

[1]*Oxford Brookes Business School, United Kingdom*

[2]*Anant National University, India*

ABSTRACT

This chapter is based on two models: the Blakeley and Visser models. The hosting study case was *Le tour des Yoles de la Martinique'*, an annual sports event in Martinique, has been explored from a tourism perspective. The chapter sought to research what specialists dubbed as "the marketability of the event" as a tourist attraction. To determine the performance of *Le tour des Yoles de la Martinique* from a performance perspective, questions related to supporting factors and resources, destination management, and qualifying determinants are to be considered. The outcomes obtained reveal that '*Le tour des Yoles de la Martinique'* is not generating a positive image per se for the destination.

7.1 INTRODUCTION

Sports event, as a tourism product, is gaining increased importance in small islands. In the context of the French overseas territories, four islands

Post-Disaster and Post-Conflict Tourism (An Updated 2nd Edition), Maximiliano E. Korstanje, Hugues Seraphin, and Vanessa G. B. Gowreesunkar (Eds.)

constitute the territory, namely French Guiana, Guadeloupe, Martinique, and Reunion Island (McElroy & Sanborn, 2005; Gay, 2012). These territories keep a common history, and they are characterized by postcolonialism, while their dependency on tourism is far from being reversed. As stated, it has been observed that the tourism industries of those French territories are stagnating or declining (Gay, 2012). This poor performance is mainly inscribed in the limited tourism activities and the poor infrastructure the site shows; for instance, event, night life and entertainment are limited, if non-existent at those French islands (Gay, 2012, pp. 1641, 1642; Dehoorne & Augier, 2011). Following the same line of observation, Ketter & Avraham (2012); and Baker (2007) point out that many worldwide declining destinations have success by boosting the local economies and their destination image by focusing on event-based marketing strategies. Beirman (2003); and Kaplanidou et al. (2013) further crystalize this aspect by stressing that the hosting of diverse types of events directed at attracting tourists is one of the preferred strategies of most island destinations. Some studies indicate that the leisure activities are key factors in the marketability of an affected destination contributing notably to the attraction of the site, which adjoins to the tourist's satisfaction (see for example, Gowreesunkar & Sotiriadis, 2015; Bhatia, 2011; Adeboye, 2012; Loi, 2008; Dwyer et al., 2004).

Martinique is a clear example of what has been said. As a declining destination, it struggles to find a right combination of indicators and factors that helps the destination in times of crisis. Séraphin (2014) argues that cultural products and services associated to the island, has been placed as a potential option to improve the tourist performance. Consequently, our hypothesis is that *'Le tour des Yoles de la Martinique,'* a popular sport event in Martinique, could be utilized in the promotion of its tourism industry. However, sports events require managers to increase the quality of the products and services they offer, to increase speed and flexibility in responding to customers, and to innovate and constantly provide new products and services' (Covell & Walker, 2013, p. 14). Constant innovation might mean that the event as it is now might use its authenticity and yet, authenticity is a major pull factor for destinations (Séraphin, 2011; Thomson, 2004). Based on these assumptions and observations, various broad questions are raised in this study: Does '*Le tour des Yoles de la Martinique*' has the potential to be promoted as a sustainable form of tourism at Martinique?

In this chapter, we are going to address two key research questions: Is it possible to turn a traditional sport event into a tourism tool without damaging the traditional aspects of the event?

7.2 TOURISM IN THE CARIBBEAN AND IN MARTINIQUE

Tourism is known by the popular opinion as a promising and growing industry (Cooper & Hall, 2008, p. 252). However, it generates material asymmetries in the Third World. As a region, which was historically associated with slavery and the colonial exploitation, the Caribbean is comprised mostly of developing countries, and has benefited tremendously from this industry (Séraphin & Nolan, 2013; Sheller, 2004, p. 23). Although Haiti and the Dominican Republic (DR) conserve different images, no less true is that insecure destinations are conceived as sites where the worst can happen anytime (Séraphin, 2011, 2014b; Higate & Henry, 2009).

The destinations of the Caribbean can be divided into three main groups: First, as the 'Big 5,' they are the destinations receiving more than 1 million visitors each year. The second group is known as the 'Followers' and they receive between half a million (or close to half a million) visitors and slightly less than 1 million visitors. The third group, the 'To be,' are between half a million visitors and 1,00,000. And finally, the 'slackers' receive less than the previous group (Table 7.1).

TABLE 7.1 Tourists in the Caribbean

Destination	Number of Visitors
Dominican Republic	4,689,770
Cuba	2,851,330
Jamaica	2,008,409
Puerto Rico	1,588,677
Bahamas	1,363,487
Aruba	9,79,256
US Virgin Is.	7,02,963
Barbados	5,08,520
Martinique	4,89,706
St. Maarten	4,67,259
Curacao	4,40,044
Haiti	4,19,736
British Virgin Islands	3,55,677

TABLE 7.1 *(Continued)*

Destination	Number of Visitors
Cayman Islands	3,45,387
St. Lucia	3,18,626
Belize	2,94,176
Turks and Caicos	2,90,587
Suriname	2,49,102
Antigua and Barbuda	2,43,932
Bermuda	2,36,343
Guyana	1,65,841
Grenada	1,16,456
St. Kitts	1,00,997
Dominica	78,277
St. Vincent and Gren	71,725
Anguilla	69,068
Montserrat	7,202

Source: Self-elaborated by authors.

According to a study by Gay (2012), Martinique, as a French Overseas Departments and Regions (DROM) has a weak performance in tourism due to the following factors (Gay, 2012):

- Lost of international appeal and subsequently their advance over their neighbors;
- Poor ration price/quality of accommodation, excursions, and restaurants;
- Poor customer service;
- Shopping is limited and expensive;
- Non-existent nightlife;
- Limited activities for tourists;
- Port and airport taxes are dissuasive;
- Lack of motivation of the people working in the industry;
- Foreign airlines have abandoned the destinations;
- Appalling social climate;
- Increase of intraregional competition;
- The EU standards limit the competitiveness of the departments and regions due to their lack of liberalism;
- Absence of international group in the hospitality sector.

However, being a non-independent destination presents some competitive advantages that independent destinations, as emerging countries do not have. For instance, good quality infrastructures; satisfactory water supply; sanitary, health conditions and emergency care are good; hospitals are well equipped and the warning and rescue systems for natural disasters are also good; they are in general safe destinations. Hence for instance one of the reasons why Martinique is so popular with French tourists. Indeed, 80% of tourists in Martinique are from France (Comite Martiniquais du Tourisme, 2009).

7.3 EVENT IN ISLAND TOURISM

According to King (1993, p. 14), islands are "the most enticing forms of land. Symbols of the eternal contest between land and water… they are detached, self-contained entities whose boundaries are obvious." Islands, as tourism destinations, have been defined from various viewpoints, but most of the definitions concur to the following characteristics: islands are small, vulnerable, and geographically insular; islands face scarcity of tourism resources with reliance on sun, sea, and sand; there is high percentage of importation; lack of opportunities on self-sustaining strategies is apparent; sustainability problems – resources, economic growth are ongoing; foreign ownership dominates tourism businesses and there is limited opportunities if not effectively marketed (*see* Gowreesunkar, 2013; Light & Dana, 2013; Brigulio, 2008; Dodds & Joppe, 2009; Dehoorne & Saffache, 2008; Pearce, 2008). These characteristics, to a large extent, reconcile with those proposed by Gay (2012) for the island of Martinique and certainly, call for specific and well-crafted marketing strategies. As a result, one of the popular workable strategies that islands choose to enhance their tourism offer is the integration of event in their respective packages. According to Avraham & Ketter (2008, 2013); and Waitt (2001), there are seven main types of event that can be workable for tourism and these have been found to offer significant advantages for the event-hosting destination:

- Mega sporting events;
- Sporting events;
- Cultural events;
- Conferences and conventions;
- Events aimed at consolidating destination brand;

- Events aimed at attracting positive perceptions from opinion leaders and celebrities; and
- Events aimed at correcting negative an element, transforming them into positive ones.

As Nielsen (2001) said, hosting sports seem to be a good option for island promotions. The point was validated and tested by countless studies and works in tourism and hospitality fields (Getz, 2008; Richards & Palmer, 2010; Raj, Walters, & Rashid, 2013) and tourism development is part of the economic impacts of events (Bladen et al., 2012; Raj et al., 2013). Events not only offer entertainment opportunities to tourists but also constitute business platforms for many related industries as well as local communities (Loi, 2009). Likewise, Avraham & Daugherty (2009) alert although sporting events are preferred options, many destinations increasingly utilize those platforms to promote local products, natural phenomena, local history or cultural traditions (Avraham & Ketter, 2013). It is important to remind that other destinations have been found to hold such events as: Summer and Winter Olympics, World Expos, music shows or concerts or competitions like the Eurovision Song Contest, as platforms that offer ample opportunity for PR-driven image campaigns on an ambitious scale. Indeed, events in tourism 'provide the means to achieve a diverse range of social outcomes, including community cohesion, educational development, support for families and regional development' (Bladen et al., 2012, p. 379). This effect of events is even stronger in destinations where events are considered as overlapping the tourism and hospitality industry and also when the idea of events as 'products' or 'attractions' (Getz, 2012) is accepted. Taking the example of Haiti, a Small Island Developing State (SIDS), Séraphin & Nolan (2013), explain that the destination would benefit from the generating effects of events. In this chapter, we are hoping to figure out the potential role that events can play on the performance of Martinique as a tourist destination.

7.4 'LE TOUR DES YOLES DE LA MARTINIQUE': A SPORT AND CULTURAL EVENT

History reveals that sports events have played a significant role in the development of society. They present the following features (Masterman, 2010):

- Vary in scale from local to global size event (hallmark, mega, major, and minor events);
- Ordinary (unplanned) or special (planned);
- Organized throughout the world;
- For able and disabled bodied men and/or women of all ages;
- Single and multi-sport formats;
- Some are universally available or specific to only one region/country;
- Competition is one day or all year-round;
- One time or recurring;
- Amateur and professional;
- Spectator or participant led;
- Corporate consumers (e.g., sponsors) or individual consumers (e.g., participants);
- Different type of owners (local government and authorities, sport governing bodies and competition organization, corporate organization, volunteer, and charitable organizations and educational institutions and organizations);
- Different participants (competitors, officials, entourage, suppliers, staff, spectators, media, VIPs).

As for '*Le tour des Yoles de la Martinique*,' it is a traditional boat race going round Martinique, with a determined starting and finishing points. It is quite a recent event, as the first race was in 1985 (Hersilie, 2001). Following the general framework provided by Masterman (2010), '*Le tour des Yoles de la Martinique*' can be described as follow:

- Local/minor event;
- Planned;
- Happens only in Martinique;
- For able bodied men and/or women;
- Competition happens only once a year during the summer over a few days;
- Recurring event since 1985;

'*Le tour des Yoles de la Martinique*,' as a small sport event, takes place on a regular basis since 1985. As events can have strong social and economic benefits for a destination with poor performance, it is plausible to propose Le '*Tour des Yoles de la Martinique*' as a potential tool for the tourism sector in Haiti. Based on the nature of this event, can it be considered as a real tourism tool? What are the assets and limitations of this sport event? The Blakeley Model is going to help to identify the limitations (and strengths) of the sport event and the Visser model, how to overcome them.

As a sport event, '*Le tour des Yoles de la Martinique*' could be assimilated to wellness tourism. Wellness tourism is a growing industry, and its demand has tripled over the decade due to the increasingly high stress level among the workforce (Douglas, 2001). Wellness tourism is generally explained as a form of tourism associated with the pursuit of maintaining or enhancing one's personal wellbeing (Andreu, Font-Barnet, & Roca, 2021). The wellness industry is multidimensional and multifaceted; it encompasses a diverse set of activities (health services, spa, beauty, fitness, personal growth, nature, and much more) and it cuts across several industries (hospitality, health, recreation, tourism among others). Wellness tourism is associated with activities that are intended to provide relaxation and stress-free stays to the tourists.

7.5 THE BLAKELEY MODEL

Evidence has accrued for more than half a century showing that, for a variety of reasons, individuals and even whole systems defend themselves against learning (Freud, 1942; Festinger, 1957; Kelly, 1963; Argryis, 1976; Vince, 2002; Blakeley, 2007b; Illeris, 2009) and the cultivation of ongoing learning requires conscious effort if it is to be sustained. There are many barriers to learning, some of which are systemic (e.g., lack of opportunity, a culture that does not support learning, lack of exposure to environmental change) and some rooted in individual psychology such as cognitive dissonance (Festinger, 1957) or Freudian defense mechanisms (Illeris, 2009). When an individual (or system) fails to learn, representations and understandings of reality no longer function effectively as changes that have taken place in the environment fail to be incorporated into the learner's mental representations of the world (Barr, Stimpert, & Huff, 1982). There is a general consensus that people defend themselves against learning when the content of that learning is in some way uncomfortable or threatening to the self. A simple example might be feedback that suggests the individual is not as competent in an area as he thought he was; rather than accept the feedback the individual might rationalize it away, suggesting that it is motivated by personal dislike on behalf of the feedback giver. Failure to learn then results in blind spots which can be defined as 'a regular tendency to repress, distort, dismiss or fail to notice information, views or ideas in a particular are that results in... [failure] to learn, change or grow in responses to changes in that area' (Blakeley, 2007b, p. 6).

In order to understand how to overcome blind spots, it is important to appreciate the process of learning as a holistic, embedded, and contextual experience (rather than limiting it to something specific such as the process of acquiring information or the acquisition of a skill). There are many models of learning in adulthood (Mezirow, 1990; Argyris, 1982; Schon, 1983) but the one we will elaborate here is Blakeley's, which places the issues of blind spots and defense mechanisms as central to the understanding of learning process (Blakeley, 2007b). Learning, according to this model, comprises four key processes: paying attention to a cue, experiencing emotions, sensemaking, and generating behaviors in a way that results in new or changed beliefs, behaviors or emotional orientations (such as increased self-confidence or openness to change). This is not a linear process: for example, it is not unlikely that emotions can be experienced first which trigger a change in behavior and it is only afterwards that one makes sense of the event and only then that one consciously notices the original cue. Furthermore, noticing a cue, experiencing emotions, making sense and generating behavior in response to the cue need not involve learning – indeed, these mechanisms simply describe the ongoing information processing characteristic of all human consciousness. Learning only results when the individual notices new and different cues or makes sense of cues in a way that generates new constructs, beliefs, emotions, or behaviors. This is best illustrated in Figure 7.1.

The inner circle represents normal information processing. When individuals pay attention to familiar cues, experience customary emotions, make sense and take action in ways that involve no change, then they are in the comfort zone, drawing on existing knowledge and skills in order to survive and achieve goals within the environment. When they pay attention to new cues, learn new emotions in relation to cues (e.g., increasing self-confidence in relation to a skill), make sense in ways that expand understanding and change behavior, they step outside of the comfort zone and into the learning zone; here they start to experience a range of emotions, depending on how novel the learning or how threatening to the self. Emotions associated with the learning zone include excitement, anxiety, frustration, anticipation,

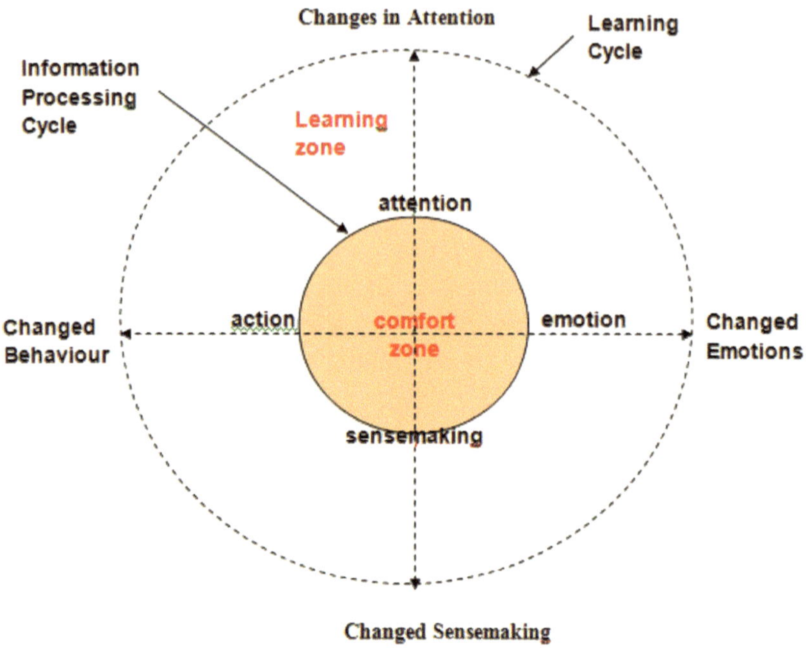

FIGURE 7.1 The Blakeley model.

Source: Blakeley (2007a).

optimism, disappointment, hope, joy, and fear. If people do not step outside their comfort zone and engage with these emotions, however, they do not learn; consequently, one of the most important skills of learning is the effective handling of the emotions. The greater the ability to do this, the greater the learning that takes place; this then contributes towards the expansion of cognitive, emotional, and behavioral resources that fit learners for the complex, changing world in which they live.

Among the motions that affect the learning we get fear, anxiety, frustration, anger, and disappointment. Any external risks that jeopardize the self-esteem creates levels of dissatisfaction which harms the image of the destination (Blakeley, 2007b; Argyris, 1982; Illeris, 2009). This then begs the question as to why might people resist lifelong learning that would appear to offer increased self-esteem and facilitate the achievement of important goals such as ongoing employment, financial security and the status that comes from holding a well-respected role in a flourishing and important economic sector?

One theory that might help to explain a resistance to learning in countries that have experienced abusive power relationships, ongoing poverty, prejudice, violence, and natural disaster, is that of learned helplessness (Seligman & Maier, 1967). Learned helplessness is a state of mind causing people to attribute their situation to uncontrollable factors such as an innate lack of ability or to global factors that will never change, making them feel powerless to change their circumstances. Research has shown that prolonged experiences of poverty or powerlessness tends to generate learned helplessness which in turn leads to a cessation in the exploration of the environment or engaging in learning related behavior (Mal, Jain, & Yadav, 1990; Teodorescu & Erev, 2014). Examples of attitudes and behaviors that might result from such a context might be:

- Nothing we do will make a difference;
- We are in this position because our people (society) are incapable;
- We are the authors of our own poverty – it's our fault;
- God punishes us for our sins;
- It's dangerous to try and change things;
- I'm safe living my life as it is.

It can be seen that the kind of learning that would need to take place would be what is termed 'double loop' (Argyris, 1976) or 'transformational' (Mezirow, 1990) which is the most difficult, rare, and most defended against as it involves a fundamental upheaval of existing beliefs and attitudes. Why would people retain these attitudes and resist learning new perspectives? There may of course be systemic challenges. For example, poverty forces people to focus on the everyday problems of ongoing family survival, leaving little energy to focus on the challenges of learning new skills and attitudes, particularly when the benefits may be distant or uncertain. However, there may also be psychological issues. It is clear, for example, that transformational or double-loop learning involves energy in 'deconstructing' attitudes and exploring their experiential and hegemonic roots (i.e., they may serve the interests of certain classes in society). It may be that continuing with existing habits and beliefs is simply the easier option. The process of believing that change is possible, investing emotional and practical energy in bringing change about, of resisting those who would oppose change or simply demonstrate indifference, requires extremely high levels of hope and optimism. The danger of raising hope is that hope is crushed, that the energy invested is wasted and that the self is seen to fail once again.

Another barrier to challenging these attitudes is the skill required to help people to recognize that the 'truth' that drives their learning is historically situated and is capable of being re-imagined. For example, if someone has performed poorly at school (due to complex systemic dynamics) then the 'truth' of their lack of competence is embodied in socially constructed symbols (e.g., school grades), roles, and attitudes. It requires sophisticated and probably facilitated processes of learning to reveal that these 'truths' need not have been the case and are in fact the result of historically contingent factors (Freire, 1996).

More practically, a belief that taking risks and setting up new businesses is worthwhile and will be supported by legal, political, social, and economic systems also involves trust. Building trust is the most difficult form of emotional learning, taking time and involving the development and deepening of relationships between people on the ground and those in power.

A common factor underlying this analysis is that the kind of learning that needs to take place is systemic and grounded in community rather than simply focused at the individual level. This poses the important question of leadership and raises questions as to whether the existing system in some way supports the interests of the ruling elite, who themselves may not be motivated to implement change.

It may be that blind spots exist at a number of levels: too much discomfort is involved in believing that things might change; beliefs such as learned helplessness are in some way comfortable and self-reinforcing, requiring high levels of energy to be challenged; the existing system of powerlessness may suit a ruling elite; a lack of leadership in facilitating profound deconstructive learning. If this in any way, describes the situation in Haiti, it seems clear that lifelong learning will only take root in the context of a planned systemic intervention led by members of the community, alongside external facilitators, with the commitment of all the key stakeholders in the tourism sector.

7.6 APPLICATION OF THE BLAKELEY MODEL TO 'LE TOUR DES YOLES DE LA MARTINIQUE'

It is necessary for Martinique to multiply its tourism products and services (Dehoorne & Augier, 2011, p. 3), creating a niche a niche that can attract new clientele (Dehoorne & Augier, 2011). Masterman (2010) argues that events have different nature and effects in publics:

- Enhanced international awareness;
- Increased economic activity;
- Enhanced facilities, services, and infrastructure;
- Increased social and cultural opportunities;
- Social regeneration;
- Land regeneration;
- Political development;
- Cultural development;
- Sport development;
- Environmental development;
- Tourism.

In the present case, the hosting of '*Le tour des Yoles de la Martinique*' is found to have limited impact (positive and negative) on the destination (Table 7.2).

TABLE 7.2 Impacts of Le Tour Des Yoles De La Martinique

Impacts (Masterman, 2010)	Le Tour Des Yoles De La Martinique Yes (Y), No (N)
Enhanced international awareness	N
Increased economic activity	N
Enhanced facilities, services, and infrastructure	N
Increased social and culture opportunities	Y
Social regeneration	N
Land regeneration	N
Political development	N
Culture development	N
Sport development	N
Environmental development	N
Tourism	N

Source: Self-elaborated by authors.

The major strength of the event relies on its capacity to stimulate social capital. For Andrews & Leopold (2013) this is absolutely important for an event.

Despite the fact there is a potential to increase backward linkages between tourism and local sport event, the right positioning is still the most important condition for the success of a destination aiming to fit into the market of the global sport industry. The first thing to do is to identify the 'blind spot' (Blakeley, 2007b), in other words, anything that can stop the destination to perform well as a (sport) tourism destination. In our case, what stops '*Le tour des Yoles de la Martinique*' to be a sustainable tool for the development of Martinique?

The Blakeley Model is based on two conditions: glocalization and globalization. As this model indicates, stepping out of the comfort zone by taking action toward a change of behavior is very important in the process of learning (Blakeley, 2007b). The organizer of the '*Tour des Yoles de la Martinique*' have international ambitions for the event but unfortunately failed to perform suitable marketing strategies to conserve the sponsorship and media coverage. In this vein, Parent & Smith-Swan (2013) argue that being at the cutting edge of technology, it is extremely important to capitalize on technology in the management of major events. At the moment, the sponsors of the '*Le tour des Yoles de la Martinique*' are small local companies. At present, the event is broadcasted by two local TV channels and is limited to the broadcast on the web only. In order to re-position '*Le tour des Yoles de la Martinique*' into a major tourism event, the Martinique DMO needs to capitalize on the event by fully embracing all the elements mentioned above, that is, sponsorship, media, and marketing (Figure 7.2).

Now that we have identified the 'blind spot,' the next step is to identify the contribution of the event to the competitiveness of the destination using the Model of Crouch and Ritchie's model (1999).

7.7 COMPETITIVENESS MODEL OF CROUCH AND RITCHIE'S MODEL (1999)

Crouch & Ritchie (1999) explain that destination competitiveness is all about the destination ability (based on resources status and utilization) to provide high quality experience to both visitors (good travel experience) and locals (better living conditions and environment). The competitiveness Model of Crouch and Ritchie's model (1999) may be both of competitive and of comparative advantage. In this study, the model is used to propose a conceptual model of '*Le tour des Yoles de la Martinique*' in terms of

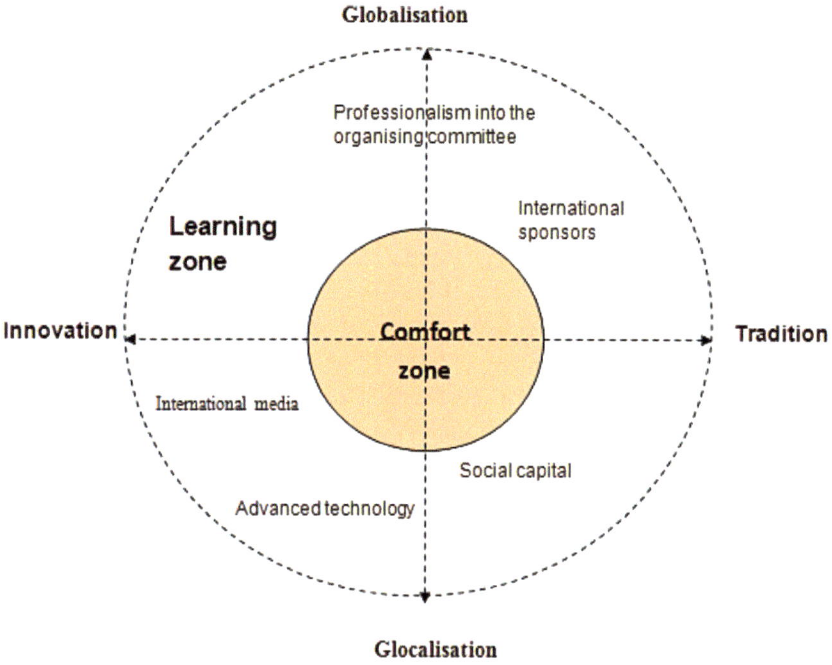

FIGURE 7.2 Blind spots in Tour des Yoles de la Martinique.

Source: Self-elaborated by authors.

destination competitiveness. It is also worth emphasizing on the fact that the Crouch and Ritchie's model (1999) is to date the best model to investigate (destination) competitiveness whether when doing qualitative or quantitative research (Koo et al., 2016). Recent research carried out by Andrades & Dimanche (2017) to discuss issues and challenges with the competitiveness of Russia as a destination also used the Crouch and Ritchie's model (1999).

In order to provide a model for '*Le tour des Yoles de la Martinique*' competitiveness, the set of questions (and approach) used by Koo et al. (2016) to determine the competitiveness of smart tourism destinations, are also used here, but adapted to '*Le tour des Yoles de la Martinique.*'

To determine '*Le tour des Yoles de la Martinique*' success and profitability, questions related to supporting factors and resources, destination management, and qualifying determinants are to be considered:

- What are the factors that can determine the success or competitiveness of *Le tour des Yoles de la Martinique*? Can they be priorities? To decide *Le tour des Yoles de la Martinique* current situation and original resources, questions related to the environment (macro and micro), core resources and attractors are to be considered:
- What are the greatest competitive strengths of existing *Le tour des Yoles de la Martinique*?
  To determine the core factors that represent *Le tour des Yoles de la Martinique*, the question asked is:
- What are the main determinants of the notion of *Le tour des Yoles de la Martinique*?
  Finally, to establish the relationship amongst core resources and attractors, supporting factors and resources, destination management, and qualifying determinants, the following question is asked:
- How does *Le tour des Yoles de la Martinique* improve their competitive position on the short and/or long term?

The factors that can determine the success and/or competitiveness of *Le tour des Yoles de la Martinique* are (in order of importance):

- The uniqueness of the event;
- The inclusive dimension of the event (gender, social background, range of stakeholders, etc.);
- The balance between modern and tradition;
- It serves many purposes as it is a mixture of social, cultural, sport, and business event.

As for the greatest competitive strengths of existing *Le tour des Yoles de la Martinique*, they are (in order of importance):

- Ability to cater for different audience/stakeholders;
- The outstanding location of the event (the Caribbean);
- The multi-purpose nature and possible use of the event.
- *Le tour des Yoles de la Martinique* could improve its competitive position on the short and/or long term by (in order of importance):
- Continuing its growth (without losing its authenticity) by adopting a more ‘aggressive’ marketing strategy;
- Continuing diversifying the types of audience it is catering for.

7.8 CONCLUSION

This chapter deals with two important models in marketing literature: the Blakeley and Visser models. The hosting study case was '*Le tour des Yoles de la Martinique*,' an annual sport event in Martinique, has been explored from a tourism perspective. We sought to research what specialists dubbed as "the marketability of the event" as a tourist attraction. The outcomes obtained reveal that though '*Le tour des Yoles de la Martinique*' is not generating positive image per se, the application of Visser and Blakeley models (preferably combined) identifies "the blind spots" that prevented the promotion of Martinique. In this respect, culturally the event-management is marginally accepted in the Island and this is the reason why further investigation is needed. The study offers a conceptual discussion revolving around the efficacy of strategy as an educational toolkit that engages the stereotypes of Western tourists to the peripheral former colonial nations. We expanded in anyway the understanding whether '*Le tour des Yoles de la Martinique*' has the potential to be promoted as a sustainable form of tourism at Martinique. The landscape of exploitation and postcolonialism, rather, still remains open for the next chapters of this book.

KEYWORDS

- **branding**
- **image**
- **island**
- **model**
- **performance**
- **sport event**

REFERENCES

Adeboye, C., (2012). *Impact of Entertainment on Tourism: Case Study, Agency Remarc in Greece.* Unpublished Doctoral Dissertation, Central Ostrobothnia University of Applied Science, Finland.

Andrades, L., & Dimanche, F., (2017). Destination competitiveness and tourism development in Russia: Issues and challenges. *Tourism Management, 62*, 360–376.

Andreu, M. G. N. L., Font-Barnet, A., & Roca, M. E., (2021). Wellness tourism — New challenges and opportunities for tourism in salou. *Sustainability, 13*, 8246. https://doi.org/10.3390/su13158246.

Andrews, H., & Leopold, T., (2013). *Events and the Social Science*. London: Routledge.

Avraham, E., & Daugherty, D., (2009). We're known for oil. But we also have watercolors, acrylics & pastels: Media strategies for marketing small cities and towns in Texas. *Cities, 26*(6).

Avraham, E., & Ketter, E., (2008). *Media Strategies for Marketing Places in Crises: Improving the Image of Cities, Countries, and Tourist Destinations*. Oxford, UK: Butterworth Heinemann.

Avraham, E., & Ketter, E., (2013). Marketing destinations with prolonged negative images: Towards a new model. *Tourism Geographies, 15*(1).

Beirman, D., (2003). *Restoring Tourism Destinations in Crisis*. Cambridge, UK: CABI.

Bhatia, A. K., (2011). *Tourism Development, Principles and Practices*. New Delhi: Sterling Publishers.

Blakeley, K., (2007a). Local governance and local democracy: The Barcelona model. *Local Government Studies, 31*(2), 149–165.

Blakeley, K., (2007b). *Leadership Blind Spots and What to Do About Them*. Chichester: Wiley.

Brigulio, L., (2008). Sustainable tourism in small island jurisdictions with special reference to Malta. *Journal of Tourism Research, 1*(1), 29–39.

Cooper, & Schindler, (1999). *Business Research Method*. New York: McGrawhill.

Cooper, C., & Hall, M., (2008). *Contemporary Tourism: An International Approach*. Elsevier: Oxford.

Crouch, G., & Ritchie, J. R. B., (1999). Tourism, competitiveness, and societal prosperity. *Journal of Business Research, 44*(3), 137–152.

Dehoorne, O., & Augier, D., (2011). Toward a new tourism policy in the French West Indies: The end of mass tourism resorts and a new policy for sustainable tourism ecotourism. *Caribbean Studies*. doi: 10.4000/etudescaribeennes.5262.

Dodds, R., & Joppe, M., (2009). The demand for, and participation in corporate social responsibility and sustainable tourism: Implications for the Caribbean. *Journal of Tourism Research, 2*(1), 1–11.

Douglas, N., (2001). Travelling for health: Spa and health resorts. In: Douglas, N., Douglas, N., & Derrett, R., (eds.), *Special Interest Tourism: Context and Cases* (pp. 260–282). Milton, Qld.: John Wiley & Sons.

Dwyer, L., Mellor, R., Livaic, Z., Deborah, E., & Chulwon, K., (2004). Attributes of tourism destination competitiveness: A factor analysis. *Tourism Analysis, 9*(1), 91–101.

Gay, J. C., (2012). Why is tourism doing poorly in overseas France? *Annals of Tourism Research, 39*(3), 1634–1652.

Getz, D., (2008). *Event Studies: Theory, Research and Policy for Planned Events*. London: Elsevier.

Gowreesunkar, G. B., & Sotiriades, M., (2015). Entertainment of tourists in leisure destinations: Evidence from the island of Mauritius. *African Journal of Hospitality, Tourism and Leisure, 4*.

Gowreesunkar, G. B., Van, D. S. J., & Séraphin, H., (2015). Social entrepreneurship as a tool for promoting global citizenship in island tourism. *Ara Journal of Tourism Research, 5*(1), 7–23.

Higate, P., & Henry, M., (2009). *Insecure Spaces*. London & New York: Zed Books.

Kaplanidou, K., & Vogt, C., (2013). The interrelationship between sport event and destination image and sport tourists' behaviors. *Journal of Sport & Tourism, 12*(3, 4).

King, R., (1993). The geographical fascination of islands. In: Lockhart, D., Drakakis-Smith, D., & Schembri, J., (eds.), *The Development Process in Small Islands States*. London: Routledge.

Light, I., & Dana, L. P., (2013). *Boundaries of Social Capital in Entrepreneurship, Entrepreneurship Theory and Practice.* doi: 10.1111/etap.12016.

Mak, A. H. N., Lumbers, M., & Eves, A., (2012). Globalization and food consumption in tourism. *Annals of Tourism Research, 39*(1), 171–196.

Masterman, G., (2010). *Strategic Sports Event Management.* London: Butterworth-Heinemann.

McElroy, J., & Sanborn, K., (2005). The propensity for defense in small Caribbean and Pacific islands. *Bank of Valetta Review*, No. 31.

Nielsen, C., (2001). *Tourism and the Media.* Melbourne: Hospitality Press.

Pearce, D. G., (2008). Tourism planning in small tropical islands: Methodological considerations and development issues in Samoa. *Caribbean Studies*. doi: 10.400/ etudescaribeennes.1393.

Raj, R., Walters, P., & Rashid, T., (2013). *Events Management: Principles & Practice*. London, Sage.

Richards, G., & Palmer, R., (2010). *Eventful Cities: Cultural Management and Urban Revitalization.* Oxford: Elsevier.

Séraphin, H., & Nolan, E., (2013). Events and tourism: The role of (Imagineering) events in local community self-actualization in small territories. The case of Haiti. *International Journal of Tourism and Hospitality* [online].

Séraphin, H., (2011). Hispaniola: The future destination of the Caribbean. *Journal of Tourism Consumption and Practice, 3*(2), 38–44.

Séraphin, H., (2014). Training: A sine-qua non condition for the sustainable development of tourism in Martinique. In: Roselé, C. P., & Raboteur, J., (eds.), *The Challenges of Tourism Development and the Economic Intelligence of Territories: Cities on Sea and on Land.* Paris: Publibook.

Séraphin, H., (2014). Visual research method approach applied to marketing to design DMO's logo. The case of Haiti. *Journal of Tourism Consumption and Practice, 6*(2), 54–58.

Sheller, M., (2004). Natural hedonism: The invention of Caribbean islands as tropical playgrounds. In: Duval, D. T., (ed.), *Tourism in the Caribbean.* Trends, Developments, Prospects London: Routledge.

Shone, A., & Parry, B., (2004). *Successful Event Management: A Practical Handbook* (2nd edn.), Londres: Thomson.

Smith, A., (2012). *Events and Urban Regeneration: The Strategic use of Events to Revitalize Cities*. London: Routledge.

Thomson, I., (2004). *Bonjour Blanc: A Journey Through Haiti*. London: Vintage.

Visser, W., (2015). *Sustainable Frontiers: Unlocking Change Through Business, Leadership and Innovation.* Sheffield: Greenleaf Publishing.

Waitt, G., (2001). The Olympic spirit and civic boosterism: The Sydney 2000 Olympics. *Tourism Geographies, 3*(3), 1–15.

CHAPTER 8

The Future of Postcolonial, Conflict, and Disaster Destinations: Vaccine Tourism in a Post-COVID-19 World

MAXIMILIANO E. KORSTANJE

Department of Economics, University of Palermo, Larrea ST 1079, Buenos Aires, Argentina

ABSTRACT

The concept of postcolonial theory has been contested by some academic circles. Anyway, we toy with the belief that the postcolonial theory is a fertile ground to understand some material asymmetries culturally enrooted in global capitalism. This chapter describes the ebbs and flows of Vaccine tourism as well as the risks posed by those travelers who flew abroad to be inoculated in the mid of the strictest lockdowns and restrictions.

8.1 THE THEORY OF DEVELOPMENT AFTER COVID-19

The world was sliced into two sides just after Harry Truman's speech dated in 1946. The event divided nations and economies into developed and underdeveloped. As Esteva puts it, the theory of development not only can be interpreted as an attempt of the westernization of non-western cultures but also represents the rise of a new dominant paradigm articulated to impose the mainstream cultural values of Occident to the rest of the world (Esteva & Prakash, 1998). From this mindset, different business organizations invested a considerable volume of loans in undeveloped

Post-Disaster and Post-Conflict Tourism (An Updated 2nd Edition), Maximiliano E. Korstanje, Hugues Seraphin, and Vanessa G. B. Gowreesunkar (Eds.)

economies (So, 1990). But despite the significant financial aid given to the Third World, no less true is that the emergent economies failed to reach a maturate state of development or at the least in the optimization of their productive means. Some critical voices alerted that beyond the theory of development laid a try to masquerade long-dormant cultural domination which was started with the colonization in former centuries. The development not only did not result as it was imagined but also created a financial dependency between the South and Global North (Stiglitz, 1989; Escobar, 2011; Korstanje, 2018). Edward Said pivoted an interesting model to describe how Europe built a narrative about what is non-West, subordinating the World to European rule. This point was brilliantly assessed by Edward Said (1979) in his book "*Orientalism*," where the author deploys several literary resources to explain how Western civilization has unilaterally created a biased image of the East to validate previously designed stereotypes and racial prejudices. Said contends that far the East was culturally invented as a place of mystery, fraught with political upheaval and inferiors, which doubtless deserved to be conquered, civilized, and subordinated to the European rule. However, Orient is something else than a simple creation, it corresponds with a dense net of narratives, sentiments, and emotional dispositions which were historically oriented to legitimate the presence of England and France in Asia. Said takes his cue from the notion of discourse which was earlier formulated by Michel Foucault. As a system of knowledge orchestrated to support the expansion of European powers, Orientalism acted not only to induce an ideological message, where the European identity was superior and ideal regarding another identity but also as an exclusionary force that was legalized by modern science. The non-western "Other" was thought of as uncivilized, less smart, or simply less virtuous than Europeans simply because there was substantial scientific evidence in the hands of scholarship. Like Orientalism, many other fictionalizations move in the collective imagination as unquestionable truths. Development, needless to say, appears not to be the exemption.

The recent COVID-19 outbreak (in 2019) subverts and interrogates the hope and goodwill intentions of Western Nations, which selfishly monopolized and reserved a greater quota of vaccines whilst leaving under-developing economies into a peripheral position. Old stereotypes and racist discourses emerged just over the colonial period that portrayed Asian tourists as the carriers of the COVID-19. Some interesting studies evince how an anti-Chinese sentiment emerged just after the COVID-19. Far from being new, this sentiment engages with the ways the Non-Western "Other" is treated in the West. This chapter not only deals with questions associated with heritage and development theory but also discusses critically the financial inequalities cemented by globalization and development programs. At the same time, the scarcity of vaccines in the Global South is a theme the section explores with accuracy. The problem of development and fair wealth distribution is brought into the foreground. Since development theory attempted to move resources to eradicate poverty, the dogma paradoxically made from it a commodity to be consumed and exchanged (Freire Medeiros, 2016).

8.2 HERITAGE AND DEVELOPMENT IN SCRUTINY

In a seminal book entitled "*Development and Social Change*" Phillip McMichael (2011) scrutinizes the intersection of development theory, at least as it was postulated in the economic circles, with the colonial legacy. The main point of entry in the discussion seems to be the role of poverty in the capitalist system as a concept that narrows to colonial domination. As McMichael observes, there was a North-South dependency that comes from the 18th and 19th centuries when Europe launched to colonize the globe. Over the years, the tug of war of colonial order was the free trade as the most valuable sign of civilization, but in this case, Europeans combined a climate of liberalism in their urban core with a state of exploitation in the hapless periphery. This allowed a successfully-orchestrated extraction of local resources which were cyclically re-elaborated in the forms of new commodities and finally reintroduced in the colony. This suggests the growth of colonies was not achieved in egalitarian terms respecting imperial powers. Although there were serious material asymmetries between North and South, no less true is that the process of decolonization in the 60s decade, which means not only the need of adopting democracy and democratic institutions by the side of peripheral colonies but by confronting European powers to gain the independence. Governance plays a leading role affirming the old center-periphery dependency, at least as it has been cemented in the European colonialism. This subordination is today replicated in the world of business and international commerce. The main thesis of this book is that Europe, by the introduction of "colonialism," established an ideological background for legitimizing their submissions to its overseas colonies for forming a vast far-flung empire. The exploitation of the non-European "Others" had a pervasive nature. Sooner, aborigines realized the double moral standards of the colonial order (McMichael, 2011).

The theory of development originally fostered some programs of investment betting Third World would be organized according to the Western rational cultural values and in that way, reaching progress and wealth. Beyond this discourse, poor nations became pourers whilst rich ones were made richer. Based on the colonial archetype of "noble Savage" which was discussed in this book, the exegetes of development alluded to cultural incompatibilities of undeveloped nations to amass wealth according to the guidelines of liberalism and capitalism. As Korstanje (2012) wrote, the ruling white elite often marks the subordinated groups but at the same time, it avoids being marked. To put this in other terms, indigenous tourism involves some aboriginal ethnicities which live in conservation parks, whereas the act of visiting Chicago city is denominated as "tourism." The label "aboriginal" not only denotes a social construction but is politically used to enhance the subordination of the marked bodies. What is important, as it was discussed in the first and second chapters of this project, tourism continued the logic of exploitation ignited by the colonial powers, through the articulation of different consumptions and heritage and patrimony.

In their book, Ethnicity Inc. Jean & John Comaroff (2009) discuss critically the role of tourism as a vehicle towards conflict instead of development. Although the concept of heritage has gained further traction over the recent years, there is a gap regarding how it is formed by the officialdom to protect their interests. The promises of a developed world came across with serious challenges in view of the civil wars, the acts of corruption which happened in Africa and other undeveloped cultures. Originally the supporters of development argued that this lack of economic maturation was directly proportional to the indifference of natives for democracy and free trade. Rather, Comaroff & Comaroff (2009) remind that from its inception anthropology emerged with the end of protecting natives from the advance of industrialism. This paternalistic viewpoint not only served colonial interests but developed an uncanny obsession for heritage preservation. Today's ethnographers witness how the industries of tourism and hospitality help local cultures which were historically relegated from the wealth distribution, to improve their economies but at a higher cost. Nation-states often impose higher taxes paving the way for a climate of tension and conflict between the aboriginal ethnicities and the executive branch. Instead of pacifying, heritage consumption and tourism not only commoditized other human beings but laid the foundations to an atmosphere of mistrust which triggered a civil war and ethnic cleansing.

Whatever the case may be, other scholars as Emmanuel J. De Kadt (1979) reasserted a mixed criticism of the theory of development that nuanced the posture of Marxist economists. Far from being a curse, development prospers in those nations that did not share a colonial legacy or was not subordinated in the past by a colonial power. After all, development is not good or bad, it adapts to the collective character of the hosting nation and of course to its history. The colonized cultures fail to be situated as developed economies because they inherit a climate of political instability that prevents poverty relief. Once again, the figure of heritage (like in de Kadt) plays a crucial role in the configuration of development in the social imaginary of the West.

Over the recent years, some interesting works stress not only the financial dependency of "Center-Periphery" but how first-world nations employ current colonies, which are tourist destinations to launder money. In a nutshell, tourism proffers to the demand of consumers an idealized paradise sometimes geographically situated in a former colony, whilst it serves to launder money coming from drug traffic and other illegal activities (Ambrosie, 2015). In this respect, Barry, Wood & Preusch (1984) call attention to the dependency of Caribbean nations which were historically adjoined to the exploitation of European nations like Spain, the United Kingdom, and France. Tourism failed to revitalize the economies of Caribbean nations like Jamaica, Haiti or Trinidad even aggravating their conditions in view of the uncontrolled foreign investors who repatriated their wealth to their native nations. The first world tourists though spent a considerable amount of purchasing, at these destinations needed to enjoy the same infrastructure they were accustomed in their homes. This not only increased the volume of imports – compared to exports – but also accelerated deficits in their trade balance. As they noted:

"*Tourism has become a trap for the Caribbean people. It has deepened the economic dependency of the region, chiefly on the United States, and has caused deep psychological and cultural damage. Caribbean tourists generally have come not to share the local culture, ambience, and cuisine, but to model the island communities in the image of themselves and their own foreign culture (Barry, Wood, & Preusch, 1986, p. 87).*"

This citation validates the assumptions of Ivan Head (1991) who claimed to reconsider the South and North relations in more egalitarian ways. Instead of seeing the South as culturally inferior or mired into a backward, we have to think about the complex interlinks between South and North, which suggests an opportunity for mutual cooperation and assistance. Like the classification of civilized and uncivilized, which were forged in former centuries, South and North are categories that are symbolically created to proclaim the superiority of one over others. This raises a more pungent question, what is the ethical role of Europe in the state of vulnerability today the Third world faces?

8.3 THE TRAUMASCAPE AND THE CONCEPT OF DISASTER-TOURISM

The turn of the century brought serious risks which placed tourism and hospitality industries in jeopardy. From the attacks of terrorists in NY, towards the rise of lethal virus or even the fury of nature in Sri Lanka or New Orleans, some specialists agree we inherited a complex world (Portroff & Neal, 1994; Shondell-Miller, 2008; Shondell-Millar & Gonzalez, 2013; Séraphin, Butcher, & Korstanje, 2017). But to some extent, the effects of these disasters in underdeveloped countries aggravates given the inequalities and vulnerabilities left by the Old Continent. The term '*Traumascape*' was initially coined by Maria Tumarkin to denote the anthropological attempt to domesticate trauma and disasters and enrooting them into a specific territory. With the benefits of hindsight, she holds the thesis that the soil embodies the meaning of events in a way that they are re-memorized after a lapse of time. Traumascape denotes more than a place of tragedy; it rather alludes to the cultural identities revolving around a tragic event created and negotiated by the involving actors (Tumarkin, 2013).

To our end, one of the most interesting studies organized to discuss the role of colonialism in the formation of vulnerability today is gazed at by first-world consumers seems to be *Thana-Tourism and Cinematic Representations of Risk*, which is authored by British sociologist Rodanthi Tzanelli. Her argumentation rests on the urgency of explaining the obsession for gazing at "other's" death as a new emerging segment of the tourism industry. Taking her cue from the concept of "tourist gaze" fleshed out by John Urry, Tzanelli explains the current inflation of risks in the media evinces no other thing than a distance between sightseers and the gazed natives. Of course, over the years, as global demand for consumption, heritage represented an elaborated fiction that took many shapes. However, heritage is changing to new forms more associated with morbid consumption. Through the sense of creativity, capitalism not only introduces a type of creative destruction but also avoid the ethical responsibilities for the exploitation of the workforce. Thousands, if not millions, of asylum seekers, refugees, and migrants knock the doors of paradise in quest of better opportunities than their homelands. Based on substantial inequalities capitalism succeeds in recycling its wastes. Tzanelli acknowledges that the concatenation of the recently occurred disasters is widely covered to entertain the public and global audiences. Through Thana-Tourism and media entertainment, the system replicates the background that facilitated the disaster. In that way, the human asymmetries are never corrected they instead are commoditized in forms of new (dark) destinations, as Tzanelli adds. Dark tourism and colonialism are inevitably entwined. The question of whether dark tourism surfaced over recent decades as one of the trends of leisure consumption in zones of extreme poverty or areas of ghettoization resulted from decades of an exploitative logic exported during the facet of colonization during the XIXth and XXth centuries. However, instead of provoking a collective awakening, cinema offers a biased diagnosis of the inequalities produced by the expansion of capitalism. It is important not to lose sight of this new global landscape, mass death seems to be a key factor in the intersection of consumption with death. This begs a more than interesting point that is not addressed by specialized literature, is the West a sadist civilization?

To answer this point, Tzanelli contends that capitalism has evolved to a new stage where artists draw a fictional landscape of "doom-days," which is centered the attention to the end of the world. The horror movies depict an apocalyptic setting where zombies strand everywhere in quest of eating fresh human flesh. This archetype represents a whole portion of rank-and-file workers who are pauperized or at least pressed to live in miserable conditions in the third world. These "exploited Others" are commodities dark tourism needs in order to pose a new version of colonialist narrative which paradoxically is tailored to the white's stereotypes. The conjuncture or environmental conditions, where movies are filmed, correspond with constituencies and representations externally designed but internally engaged by in-group members. At the periphery, nations accept an allegory that explains further their so-called state of under-development, but in blaming others, central nations face an atmosphere of terror given by the internal enemy. Central economies have serious problems controlling forced migration not only because they feel the enemies of democracy are within, but also because they have developed a radical image of the world, as an uncivilized and hostile place that deserves to be disciplined. Not surprisingly, the genealogies of District 9 in ghettoized South Africa offer a good explanation for this. There is a type of romanticism behind colonialism which dictated the future of its former colonies. What is more than important to debate, colonial powers, as well as authorities, fabricated and circulated certain narratives and stereotypes which have been not only internalized by locals, but also made of their life a good story to be told. But in so doing, the real reasons by means colonialism expanded and African cultures were ushered in pauperism are cynically covered. The fact is that dark consumption exhibits a fertile ground for capital owners to usufruct with human pain engendering a representational (but not for this real) discourse, which is aimed at legitimating the colonial past. Dark tourism not only refreshes the needs of belonging to democratic cultures which are blessed by the auspices of capitalism but also blame native for the arbitrariness and crimes committed by the European states in the colonialism. As Tzanelli puts it, the question of whether tourists are altruistically obsessed to be in contact with the pour others to feel like them is not resolving the ethical liabilities of their states in the complicated situation of the Third World (Tzanelli, 2016). To what extent may Europeans help others without facing serious charges for the omissions or crimes of their ancestors?

In her book "*Heritage that Hurts*," Joy Sather Wagstaff (2016) carries on the discussion above noted. It is important not to lose sight of much work that should have to be done to expand the current understanding of dark tourism and heritage. Centered her study case in Ground Zero, NY and 9/11, Sather Wagstaff explores the allegories around dark consumption, not as a form of commoditization but as a much deeper expression of solidarity. One of the feelings of pride of Western civilization and the US, the World Trade Center was suddenly targeted by two civilian airplanes. This coordinated attack revealed two important aspects of politics. At a first glimpse, the military supremacy of the US was vulnerated weaponizing some commercial airplanes. This shocked not only Americans but the rest of the world. Secondly and most importantly, countless governments and leaders unilaterally expressed their solidarity with the United States and its people. As she recognizes, dark tourism acts as a conduit towards the respect of the alterity, mediating between the present and the future. Sather Wagstaff seems to be seriously affected by September 11. She argues convincingly that the specialized literature does not understand that dark tourism opens the doors to express human solidarity, producing emotional performance and shared experiences even in zones far away from the US. No matter the degree of authenticity, dark tourism places all men in egalitarian conditions before death. Visitors not only imagine their finitude but also they are in quest of answers that in help understanding the complexity of this world. In some cases, as she adheres, the political power intervenes in the formation of the dark site, manipulating politically the message to tell. This is exactly the best example of the heritage, which far from being real, is distorted for the ruling elite to preserve their status quo (Sather Wagstaff, 2016). Phillip Stone reminds us that the essence of dark tourism is thanatopsis, which interprets the need of figuring the own death through the lens of others. What visitors look, is not only being closer to the victims but pinning their own life according to the human suffering (Stone, 2012).

Last but not least, in a recent publication, Korstanje (2016) coined the term '*Thana-Capitalism*' to lump many derived terms and meanings as Thana or Dark tourism, Slum-Tourism, Disaster tourism or post-conflict tourism together. No matter the shape, after 9/11, he avers, a new dark tendency emerged in the core of tourism consumption. The lifestyles and patterns of consumption our gran parents enjoyed set the pace to new morbid forms. The rise of dark tourism evinces not only a novel travel trend but it represents a sign of a much deeper-seated issue, the obsession for other's death. Per Korstanje, the process of secularization from a historical perspective not only undermined the hegemony of the Catholic Church but also declined the influence of religion in daily life. As a result of this, the kingdom of Heaven was irreversibly replaced by the Kingdom of the earth. People do not believe in God or in the afterlife. Whilst medieval peasants lived shortly with the hope to enter at Paradise later, modern citizens are extremely frightened of death. By gazing at how others' died, modern consumers reinforce their aura as outstanding people, as special in a trace where death has a bad connotation. Korstanje concludes that Thana

Capitalism is based on a climate of extreme competence, comparable to social Darwinist grounds, where the death of others is situated as the main commodity to exchange. To some extent, this does not resolve if dark tourism or post-disaster marketing can be used as instruments of resilience in post-disaster contexts.

8.4 REFLECTIONS AROUND POST-DISASTER MARKETING

Post-disaster tourism or marketing probed to be efficient toolkits to help communities in the recovery timeframe. Many examples ranging from New Orleans to Sri Lanka validate the assumption although tourism seems to be highly vulnerable to external and major threats, no less true is that it works as a mechanism of resilience organizing the post-disaster recovery steps (Robinson & Jarvie, 2008; Beirman, 2003; Walters & Mair, 2012; Scott, Laws, & Prideaux, 2008; Korstanje & Ivanov, 2012). Scholars of different stripes devoted considerable attention and efforts to anticipating disasters to avoid their negative effects. However, recently and because of risk perception theory to predict the future, a whole portion of colleagues toyed with the idea of post-disaster marketing fits in policymakers to construct more resilient and safer destinations as well as mitigating the devastation in post-disaster landscapes (Laws, Prideaux, & Chon, 2006; Visser & Ferreira, 2013; Tzanelli, 2016; Korstanje, 2016). But the matter did not rest there, as Zygmunt Bauman (2013) eloquently alerted, the current atmosphere of fear and dangers, which is amply stimulated by mass media may very well lead to a paradoxical situation. On one hand, it is creating a precautionary platform that keeps the audience under control whilst on another, the risks of a major disaster takes a hit turns higher. The degree of insensibility produced by the bombarding of risk-related news is paving the way for a major disaster to happen. This was exactly what French Philosopher Jean Baudrillard dubbed as "the Spectacle of Disaster." He holds the thesis that the nature of hyper-reality does consists in interrogating the present from the future, but since the future never takes shape, we in the present are moved by pseudo-events, which means events that never happen (Baudrillard, 2005, 2006). Baudrillard's insight into dark tourism issues, though less recognized by professional researchers and field workers, is vital because of two main reasons. Whilst the government takes a direct intervention to thematize the disasters, it lays the foundations to distort its causes and reasons. As a result, the probabilities of the same event repeats are multiplicated. Paradoxically, we are prone to theatricalize disasters but once it is done, a new disaster emerges again. Secondly, the Spectacle of Disaster offers a simulacrum that impedes correcting the factors that ushered the community into the tragedy. Because of its complexity and profundity, dark tourism site sells only a part of the history. Dark tourism tells a partial truth revolving around the crimes in Nazi Germany or Ground Zero in which case, a similar event can happen again, again, and again!. To set an example, over the latest elections the third party was the Alt-right a new ally of right-wing extremists, White-Supremacists, Racists, and Neo-Nazis than overtly declared as staunch enemies of democracy, capitalism, and mass migration. Equally important, Donald Trump reached the American presidency stimulating a debate against immigrants and the Muslim community. The lesson is what did we learn from past events?

Of course, the thoughts struck me that they never fought in WWII or were directly involved in the reconstruction of Berlin, to think Adolf Hitler (and others messianic maniacs) would be an option to prevent Muslim-related terrorism. As David Altheide (2017) brilliantly hinted, terrorism reminds not only the powerful influence of fear-mongering but how one fear is replaced by another. The latest presidential debate between Hilary Clinton and Donald Trump evinced the rise of two contrasting fears, which were historically encapsulated in the core of the United States. Whilst Clinton supporters emphasized the risks of tyranny and the subsequent decline of democracy – if white supremacists win-, Trump electorate energetically invoked a long-dormant fear of strangers using the articulation of migration bans and the imperative to make America Great again, which was prompted by the climate of distrust inaugurated by terrorism – just after 9/11. Altheide concentrates on outlining an innovative model to understand how media commoditizes terrorism to entertain the public. As Altheide puts it, fear has biological roots, more oriented to the survival of subject, over recent years mass media developed a direct line for the lay citizens to get informed not only about what is happening elsewhere but on the external threats that may very well place the American society in jeopardy. In any case, one of the ethical risks in providing people with the information they are unable to grasp seems to be associated with the strange symbiosis between profits and fear, a moot point that merits to be discussed. Whereas media gained further profits by disseminating terrorism-related news, ISIS took advantage of such publicity to place its Horse of Troy in western cities. However, Altheide acknowledges that the climate of fear does not derive directly from terrorism. It seems to be an all-pervading force socially constructed and unfolded to keep the work-force under control in the capitalist society. In this way, the second chapter explores critically how fear is manipulated by top-ranked politicians to introduce substantial changes in the already-existent laws and regulations regarding surveillance and homeland security. A further approach on what he named as "the Politics of Fear" suggests not only the passive acceptance of new practices, which otherwise would be neglected, but the formation of an internal atmosphere where discursively terrorism is equaled to risks of any other nature such as drug addiction, crime, or the doctrine of the precautionary principle. Cass Sunstein (2005) elicits an interesting critique on the precautionary principle as the source of new populists claims which alter the functioning of consolidated democracies. The imposition of risks not only paralyzes the masses but activates more irrational responses that can worsen the trouble. Originally, Sunstein's worries relate to knowing furtherly why laypeople are frightened or at the best, why they feel more vulnerable than other times. In deliberative democracies, the reasons for risks are discussed among the different agents at stake, but this does not happen in the dictatorship, where decisions are made at the top of the political hierarchy. However, even in a democracy, luring populism may obscure the reasons of risks moving the community into a tragedy. In other words, the state of a disaster that involves a community might be prevented or partly mitigated whenever the issues that impinge on public life are previously discussed, debated, and forecasted. This belief would explain the reasons why democratic societies have more instruments to face disasters than totalitarian or authoritarian ones. Whereas the latter does not provide their citizens with the necessary steps to evaluate the pre-existing risks, the former invests a considerable amount of capital in the process of mitigation and preparedness for natural catastrophes.

The problem of public fear is inextricably interrelated to the extent potential hazards are censored by the state. Sunstein goes on to acknowledge "democracies do best if they abstract from the largest questions and try to obtain a consensus from people who disagree on, or are unsure about, how to resolve those questions. In the context of fear, I suggest, it is possible to obtain just such a consensus… I understand fear to depend on some kind of judgment that we are in danger (Sunstein, 2005, p. 3)."

Starting from the premise that fears are pre-determined by previous beliefs rooted in the ways of perceiving disrupting events, people seem to be prone to experience serious risks in matters that are relatively safe, whereas in some circumstances avoid real threats because of ignorance or even imprudence. His main thesis is that the precautionary principle stems from a focal belief about health, safety, and environment born in the core of European enlightenment during the XVIIIth and XIXth centuries. Nonetheless, fear like other underlying emotions works like a cascade or better a virus that contaminates other hosts. In perspective, the social day-to-day interaction determines what can be considered a frightful event.

Sunstein criticizes the thesis that points out that Europe accepts the precautionary principle whilst the United States refuses it. Since Europeans overtly admit that the world should be contemplated within a margin of risk, there will be some gaps at times when capitalist states take the necessary steps to secure the life of their citizens. Conversely, American society seems to be unconcerned regarding the risks of global warming or genetic food modification policies. This suggests an erroneous idea that the United States requests proof (evidence) of potential dangers to take action. A false opposition between the United States and Europe leads scholars to a misleading direction in their reflection about the impacts of fear.

With this background in mind, Sunstein addresses an objective definition of risk linked to the potential danger a person can suffer, or even certain probabilities a subject physically or psychologically can be injured by a third party. In other words, the risk should be adequately verified and tested by scientific-based research. Risks are often linked to empirical social practices of laypeople may very well affect the lives of others. Practical policies or strategies can be discouraged or encouraged depending on the degree of risk they represent for US soil. As discussed, the idea of the end of the world is captivating a wider audience and public as never before. Since the world appeared to be created our cosmology demands, it seems to be destroyed one day. To expand our current understanding of post-disaster marketing, we need to delve into the theory of Bottoms days, at least to grasp why it remains still a captivating theme for hyper-globalized audiences and lay-people.

The recent outbreak of COVID-19 has certainly revived the discussion on the utility of post-disaster marketing to be applied to recover the tourism industry. For example, vaccine tourism offers a fertile ground not only for future applied research but to stimulate depressed economies.

The US, over the recent months, have situated as a privileged destination of vaccine tourism. Thousands of foreign tourists select the US to be inoculated. This emerging sub segment is moved by the lack of vaccines in their societies. Some voices have alerted on the risks that vaccine should be globally distributed. Some rich nations have hoarded a great portion of available vaccines leaving to undeveloped economies in a state of economic crisis. What is more important, vaccine tourism opens the doors to partisan discussions revolving around the real motivations of tourists. Whilst the rich nations are ideologically presented as civilized societies open to foreign tourists, at the bottom they emulate the core of the manifest inequalities between developed and underdeveloped nations. Higher purchasing power segments and high-class tourists travel abroad to be vaccinated, whereas the rest of the population keeps immobile.

8.5 VACCINE TOURISM AND THE FINANCIAL INEQUALITIES

The term pandemic derives from the presence of infectious disease, likely associated with a virus that cannot be contained by health authorities; this virus rapidly expands to other countries and continents, affecting the global trade, and of course the tourism industry. In the constellations of tourism, the literature says that tourism is the main carrier for virus outbreaks worldwide simply because of the recent technological breakthrough in mobilities. It is important to discuss the long-lasting effects of COVID-19 in tourism and global trade today. Some voices have emphasized the opportunity opened by COVID-19 to embrace a more sustainable tourism industry (Gossling, Scott, & Hall, 2020; Zenker & Kock, 2020; Higgins, 2020). In this respect, other studies alert on the importance to reconsider the methodological problems of tourism research, precisely in a world without tourists (Zheng, Goh, & Wen, 2020; Wen et al., 2020). Digital technologies occupy a central position not only in travel behavior but also in safety and security issues in a post-COVID-19 world (Gretzel et al., 2020). Aside from this, on 20 March of 2020, Argentinean President Alberto Fernandez decided to implement a lockdown closing border and airspace, a measure that led to canceling international and domestic flights. As a result of this, thousands of Argentinean tourists were stranded abroad without financial assistance, food or lodging. The popular opinion pressed the Ministry of Foreign Affairs to head a program of repatriation whose costs were fully absorbed by the government. These measures were mainly oriented to protect tourists as well as public health. However, some tourists even deifying the presidential decree fled abroad in the middle of the lockdown. In 2021, to be precise one year later, other tourists traveled to the US to be inoculated because of the lack of vaccines in Argentina. On 26 June of 2021, President Fernandez passed a new resolution – probably in consonance with Migratory Office—to reduce the quotas of Argentinean tourists who can come back to the country; this quota originally was set in 2000 persons was cut to 600 passengers per day. In consequence, thousands of flights were canceled or delayed whilst tourists were ultimately stranded in the US and Europe. Unlike the first expatriations, in this case, tourists were never assisted. Arguing the need of avoiding Delta strain, health authorities applauded enthusiastically the resolution. The resolution in discussion caused some panic and confusion in popular opinion because thousands of Argentinean tourists were left abroad without protection. Tourists alluded to a punishment inflicted by the government because they traveled to Miami to be vaccinated. Vaccine tourism evinces two important aspects of the pandemic. On one hand, the ideological (or partisan) discourses against the rich countries that monopolized a great portion of vaccines leaving the peripheral nations adrift. On another, the pandemic showed clearly the material asymmetries between having and have-not nations; a moot-point development theory failed to resolve. The COVID-19 accelerates a much deep crisis in service sectors aggravated by the implication of countless social maladies such as racist expressions, separatism, hostilities against Asian tourists, as well as long-dormant geopolitical tensions (Gowreensunkar et al., 2020). As Francesc Romagosa (2020) observes, the pandemic has led global capitalism into an unparalleled state of confusion and crisis. No matter their ideological background, specialists have no firm grounds for how much the crisis lasts. The tourist-phobia seems to be one of the remarkable aspects and problems left by COVID-19. People are traveling adopting domestic tourism as a primary form of consumption. Korstanje (2020) used the term '*wicked gaze*' to symbolize a new stage of feudalization (fragmentation) where tourists are considered undesired guests. The hyper-globalized world, after COVID-19, sets the pace to an atomized landscape characterized by the rise of recalcitrant nationalism. Tourists who were seen as ambassadors of prosperous and democratic societies are now scrutinized and demonized as carriers of a lethal virus. To put the same in bluntly, political tensions are one of the quintessential features of the post-COVID-19 world.

As the previous argument is given, the present section shows the preliminary results of exploratory research conducted over a sample of 25 Argentinean tourists stranded in the US. Although the outcomes are promising, they cannot be extrapolated to other universes because the method sampling was not statistically representative. Having said this, we used the snowball method which means that each interviewee recommends the next person to contact. Outcomes obtained in this research contribute notably to the applied research and the literature whereas laying the foundations towards a new understanding of the COVID-19 effects in the tourism industry.

Interviewees manifested a strong sentiment of discontent and apathy against President Fernandez. They travel to the US and Europe not only for tourism but also to be inoculated, given the scarcity of vaccines in Argentina. In many cases, they feel frustrated and abandoned by their government. Marcelo (male, 28 years old) claims:

> *"I am stranded in the US without financial support and food. I am living with what some friends give me. I feel abandoned by the government, by the state. I have never thought President Fernandez makes a disposition like this. I am thousands of miles away from my home, my family ... fully disappointed of this! If you ask, I do believe the government wants to punish us because we opt for being vaccinated in the US."*

This coincides with Marcela (female, 35 years old) who overtly says:

> *"I am calling to the Aerolineas Argentinas [Argentinean Airlines] since this morning and my flight was scheduled to August 4. I called several times to change that but without any result. I am desperate! I need to come back urgently. Like me, almost 200 compatriots are sleeping everywhere at the airport... this is a clear implicit sanction of this communist government, we live like in Cuba people flies but cannot return."*

These tourists said that the government punished them because they traveled to the US which is seen as the cradle of corruption and imperialism. Interviewees allude to the government keep a partisan (left) ideology that marginalizes them. These tourists feel discriminated against by Fernandez' administration. The crisis provoked by COVID divided public opinion into two bands: those who are in support of health measures adopted by Fernandez and those who see Fernandez undermines their basic rights. For the former, tourists are treated as hedonist travelers who maximize their pleasure even putting society at risk. In different circumstances, President Fernandez and his ministers directed their criticism against tourists who violated the quarantine or kept insensitive to the current sanitary dispositions. Juan (male, 59 years old) exclaimed:

> *"I am annoyed with this government. They treat us as citizens of the second class. The government discriminates against us thinking we are leisure travelers who are part of a privileged elite. I was working in the US when my work visa expired. Now I need to come back to Argentina. The Government closes the doors to us because they hate us."*

Maria (female 22 years old) is stranded in London given the fact that she was a postgraduate student. She replies:

> *"This pandemic divides the world in two, those who can travel (like us) and those who lost their jobs (because of the failures of the governments to stop the pandemic). This government retaliates on the mobile citizens who can travel abroad. This is more complex than hating it is a new state of the world where tourists are demonized as potential evildoers or carriers of the COVID-19."*

In both interviews, participants manifest government undermines them as second-class citizens, marginalizing them into a peripheral position. In part, the government is still insensitive to their needs for assistance abroad.

It is very interesting to see how other stories are oriented to evince a manifest fear of traveling after the pandemic. In opposition to these above-described interviewees, they take part of a minority that exhibits a much deeper sentiment of panic to traveling in the future. This seems to be the case of Jose (male 35 years old) who says:

> *"I do not believe that Fernandez should be blamed for this. The problem is that I have a manifest panic to travel after the pandemic. It looks like I will never travel again in the future. What I feel now, is the same I felt after 9/11 (the attacks on the World Trade Center). I was almost four*

> *years without flying... now I feel the same sentiment of panic in my veins and body."*

Similarly, Rose (female, 45 years old) casts some doubts on the possibility to start a leisure journey in the future:

> *"I planned this trip to the US over months I know it is not the best time. We are facing a pandemic! I suffered agoraphobia in the past for what my therapist suggested to me to make this travel. I do not know when my flight will be placed. I am desperate without money and lodging, I am sleeping at the airport."*

This suggests that events like 9/11 or the COVID crisis activate some traumatic experiences which paralyze travelers. This sentiment of aversion seems to remain stable over years until other similar traumatic event takes place.

8.6 CONCLUSION

Doubtless, the history of development theory and colonialism are inextricably intertwined. Ideologically speaking, the colonial expansion was successfully orchestrated according to two primordial paradigms. On one hand, the European law was imposed not only as of the only rule for domesticating the local, as an ideal to be exported to the world. On another, the European traveler was symbolically situated as a privileged citizen whose experiences were fruitful to be documented. The novel writing as well as other traveling novels resulted from the sentiment of supremacy marked by European Nations. The collapse of European Empires opened the ways for the rise of a new discourse oriented to divide the world in two: the development theory. Just after Truman's discourse, the world was split into two parts: developed vs. Underdeveloped nations. Per Truman's paternalism, developed nations are morally obliged to re-educate and give financial assistance to underdeveloped ones. Of course, like colonial paternalism, the development theory just failed to promote fairer wealth distribution. The fear of the "Non-Western Other" occupies a central position in the configuration of Western nations. What is more important, this fear was notably potentiated with the turn of the XXth century and the so-called War on Terror. The idea of living with the enemy inside was a quintessential feature of what D. Altheide dubbed as "the culture of fear." The outbreak of COVID-19 far from solving the gap between have and have-not economies, as well as the fear of the "Other," strengthened the anti-foreigner sentiment, now so to speak directed against Asian Tourists. The present chapter explored not only the problem of Alterity in the Western cultural matrix but also philosophical quandaries beyond vaccine tourism. For the status quo, democratic nations have developed further resources which allowed rapid acquisition of vaccines in comparison with undemocratic nations. But at a closer look, this speaks less of the inequalities between rich and poor countries.

KEYWORDS

- **COVID-19**
- **global asymmetries**
- **inequalities**
- **lockdown**
- **pandemic**

REFERENCES

Altheide, D., (2017). *Terrorism and the Politics of Fear.* New York, NY: Rowman & Littlefield.

Ambrosie, L., (2015). *Sun and Sea Tourism: Fantasy and Finance of All-Inclusive Industry.* Newscatle upon Tyne: Cambridge Scholar Publishing.

Barry, T., Wood, B., & Preusch, D., (1984). *The Other Side of Paradise: Foreign Control in the Caribbean.* New York: Grove Press.

Baudrillard, J., (2005). Violence of the virtual and integral reality. *International Journal of Baudrillard Studies, 2*(2), 1–16.

Baudrillard, J., (2006). Virtuality and events: The hell of power. *Baudrillard Studies, 3*(2), 1–9.

Bauman, Z., (2013). *Liquid Fear.* New York, NY: John Wiley & Sons.

Beirman, D., (2003). Restoring tourism destinations in crisis: A strategic marketing approach. *CAUTHE 2003: Riding the Wave of Tourism and Hospitality Research*, 1146.

Comaroff, J., & Comaroff, J., (2012). *Ethnicity Inc.* Chicago, IL: Chicago University Press.

De Kadt, E., (1984). Tourism: Passport to development. *Perspectives on the Social and Cultural Effects of Tourism in Developing Countries*. Oxford: Oxford University Press.

Escobar, A., (2011). *Encountering Development: The Making and Unmaking of the Third World*. Princeton, NJ: Princeton University Press.

Esteva, G., & Prakash, M. S., (1998). Beyond development, what? *Development in Practice, 8*(3), 280–296.

Faulkner, B., & Vikulov, S., (2001). Katherine, washed out one day, back on track the next: A post-mortem of a tourism disaster. *Tourism Management, 22*(4), 331–344.

Freire-Medeiros, B., (2014). *Touring Poverty*. Abingdon: Routledge.

Gössling, S., Scott, D., & Hall, C. M., (2020). Pandemics, tourism and global change: A rapid assessment of COVID-19. *Journal of Sustainable Tourism*, 1–20.

Gowreesukar, V., Maingi, S., Roy, H., & Micera, R., (2021). Introduction: Destination management solution post pandemic: A need of the hour! In: Gowreesukar, V., Maingi, S., Roy, H., & Micera, R., (eds.), *Tourism Destination Management in a Post Pandemic Context* (pp. 1–10). Wagon lane, Emerald Group.

Gretzel, U., Fuchs, M., Baggio, R., Hoepken, W., Law, R., Neidhardt, J., & Xiang, Z., (2020). e-Tourism beyond COVID-19: A call for transformative research. *Information Technology & Tourism*, 1.

Head, I., (1991). *On a Hinge of History: The Mutual Vulnerability of South and North*. Toronto: University of Toronto Press.

Higgins-Desbiolles, F., (2020). Socializing tourism for social and ecological justice after COVID-19. *Tourism Geographies*, 1–14.

Korstanje, M. E., & Ivanov, S. H., (2012). Tourism as a form of new psychological resilience: The inception of dark tourism. *Cultur, 6*(4), 56–41.

Korstanje, M. E., (2012). Reconsidering cultural tourism: An anthropologist's perspective. *Journal of Heritage Tourism, 7*(2), 179–184.

Korstanje, M. E., (2016). *The Rise of Thana Capitalism and Tourism*. Abingdon: Routledge.

Korstanje, M. E., (2018). *The Global Paradox: A Critical Analysis*. Cheltenham: Edward Elgar.

Korstanje, M. E., (2020). Passage from the tourist gaze to the wicked gaze: A case study on COVID-19 with special reference to Argentina. In: George, B., & Mahar, Q., (eds.), *International Case Studies in the Management of Disasters* (pp. 197–212). Wagon Lane, Emerald Publishing Limited.

Laws, E., Prideaux, B., & Chon, K. S., (2006). *Crisis Management in Tourism*. Wallingford: CABI.

McMichael, P., (2011). *Development and Social Change: A Global Perspective: A Global Perspective*. London: Sage Publications.

Pottorff, S. M., & Neal, D. D. M., (1994). Marketing implications for post-disaster tourism destinations. *Journal of Travel & Tourism Marketing, 3*(1), 115–122.

Robinson, L., & Jarvie, J. K., (2008). Post-disaster community tourism recovery: The tsunami and Arugam Bay, Sri Lanka. *Disasters, 32*(4), 631–645.

Romagosa, F., (2020). The COVID-19 crisis: Opportunities for sustainable and proximity tourism. *Tourism Geographies, 22*(3), 690–694.

Said, E., (1979). *Orientalism: Western Representations of Orient*. New York, NY: Vintage.

Sather-Wagstaff, J., (2016). *Heritage that Hurts: Tourists in the Memoryscapes of September 11*. Abingdon, Routledge.

Scott, N., Laws, E., & Prideaux, B., (2008). Tourism crises and marketing recovery strategies. *Journal of Travel & Tourism Marketing, 23*(2–4), 1–13.

Séraphin, H., Butcher, J., & Korstanje, M., (2017). Challenging the negative images of Haiti at a pre-visit stage using visual online learning materials. *Journal of Policy Research in Tourism, Leisure and Events, 9*(2), 169–181.

Shondell, M. D., & Gonzalez, C., (2013). When death is the destination: The business of death tourism–despite legal and social implications. *International Journal of Culture, Tourism and Hospitality Research, 7*(3), 293–306.

Shondell, M. D., (2008). Disaster tourism and disaster landscape attractions after Hurricane Katrina: An auto-ethnographic journey. *International Journal of Culture, Tourism and Hospitality Research, 2*(2), 115–131.

So, A. Y., (1990). *Social Change and Development: Modernization, Dependency and World-System Theories* (No. 178). London: Sage.

Stiglitz, J. E., (1989). Markets, market failures, and development. *The American Economic Review, 79*(2), 197–203.

Stone, P. R., (2012). Dark tourism and significant other death: Towards a model of mortality mediation. *Annals of Tourism Research, 39*(3), 1565–1587.

Sunstein, C. R., (2005). *Laws of Fear: Beyond the Precautionary Principle* (Vol. 6). Cambridge: Cambridge University Press.

Tumarkin, M., (2013). *Traumascapes: The Power and Fate of Places Transformed by Tragedy*. Melbourne: Melbourne University Publishing.

Tzanelli, R., (2016). *Thanatourism and Cinematic Representations of Risk: Screening the End of Tourism* (Vol. 176). Abingdon: Routledge.

Visser, G., & Ferreira, S., (2013). *Tourism and Crisis*. Abingdon: Routledge.

Walters, G., & Mair, J., (2012). The effectiveness of post-disaster recovery marketing messages—The case of the 2009 Australian bushfires. *Journal of Travel & Tourism Marketing, 29*(1), 87–103.

Wen, J., Wang, W., Kozak, M., Liu, X., & Hou, H., (2020). Many brains are better than one: The importance of interdisciplinary studies on COVID-19 in and beyond tourism. *Tourism Recreation Research*, 1–4.

Zenker, S., & Kock, F., (2020). The coronavirus pandemic–A critical discussion of a tourism research agenda. *Tourism Management, 81*, 104164.

Zheng, Y., Goh, E., & Wen, J., (2020). The effects of misleading media reports about COVID-19 on Chinese tourists' mental health: A perspective article. *Anatolia, 31*(2), 337–340.

CHAPTER 9

Food Events and Destination Image: A Corollary Note

HUGUES SÉRAPHIN[1] and VANESSA G. B. GOWREESUNKAR[2]

[1]*Oxford Brookes Business School, United Kingdom*

[2]*Anant National University, India*

ABSTRACT

This chapter reviews the significant interplay between food events and national identity. Beyond the food industry, as the authors note, there is a type of commoditization that offers a good opportunity for the local economy, but this mechanization of consumption leads to a state of depersonalization and discomfort for some visitors. In a post-COVID-19 world, the knowledge of gastronomy converges with what authors call "ambidextrous marketing," which takes the lead in managing contrasting contexts.

9.1 TOURISM DEVELOPMENT, GASTRONOMY, AND IDENTITY: WHAT KIND OF RELATIONSHIP?

Nutrition is constantly changing, and their transformations also affect society. Tourism has become part of this phenomenon. Progress in communications and transport technology have radically altered the speed with food and food information which are transmitted around the world (Richards, 2002). Ritzer (1993) speaks of "industrial cooking" with

Post-Disaster and Post-Conflict Tourism (An Updated 2nd Edition), Maximiliano E. Korstanje, Hugues Seraphin, and Vanessa G. B. Gowreesunkar (Eds.)

respect to the progress that is being made in transport (transport, railways, and cars), food production, and storage and retail, all of them have effects on what we eat. This context leads to the McDonaldization (Ritzer, 1993). Tourism pressure inevitably produces crowding out effects, even cultural type (Adams & Parmenter, 1995; McCool & Martin, 1994). The local economy attracted by the possibilities of rapid earning is available to lose its identity. For example, the contribution of Dong et al. (2003, p. 168) illustrates the behavior of the tourist destination of Dali, which is in the development stage: "An interesting finding was that most of Bai people now think of their culture as a commodity and are willing to change traditions to fit the market needs – an example is their desire to change the way they serve food from banquet style to western style due to greater efficiency."

The relationship between tourism and food within the development of a tourist destination is not stable and fixed but changing in terms of tourism development conditions. As already mentioned, excessive pressure on tourism changes the identity of a place, transforming it into the use and consumption of mass tourism with a scenario of "Disneyification" in which "pasts are condensed into easily consumed bite-sized pieces lacking any authenticity" (Ashworth, 2009, p. 80). In fact, this situation corresponds to the top of a process that begins with the development of a tourist destination. If we refer to the Butler model and two of the best-known tourist ratings, namely Plog and Cohen, it is possible to verify how the attractiveness of the gastronomy undergoes, over time, the negative externalities of crowding out typical of tourism in the consolidated phase.

In the Butler model (1980), gastronomy is indeed a very strong attraction in the early stages of growth. Several authors (Kruczek & Szromek, 2011) argue that gastronomy is decisive in the involvement phase: people from the region start raising revenue from gastronomy or other services. This is the period when the conveniences for tourists are welcome the most. At this stage tourists are characterized to be interested in elements linked to the identity of the geographic location. Plog defines such tourists as allocentric, while Cohen calls them drifters. They are both looking for integration with the host community thanks to the knowledge of the gastronomy. Fields (2002) state that finding that "hidden" little restaurant patronized by "locals" is also a goal for many tourists. This type of behavior can create a paradox, however, the status can only be gained from the experience, if others know about this kind of consumption/location, then it will be possible to take the risk, that the 'hidden' gastronomic treasure could destroy its distinctiveness.

Cohen (1972, p. 168) states that "he tries to live the way he visits live, and to share his shelter, foods, and habits." Only at this stage, therefore, gastronomy can preserve its identity characteristics. Subsequently, the destination undergoes the influence of the characteristics of the changing demand. During the consolidation and then stagnation there are tourists psychocentric (Plog, 1974) or as defined by Cohen (1972) organized mass tourists. They want to experience a tourist experience without risk, what Cohen calls "Tourist Bubble."

This situation is obviously intended to change the increase in tourist pressure. As pointed out by Garay & Cánoves (2011) describe tourism development in Catalonia, they state that after the initial exploration phase a new phase of involvement has begun in which local stakeholders have started to professionalize and industrialize the supply of accommodation and restoration: "if in the previous proto-tourism stage, hotels, and restaurants had still maintained a traditional kind of production, with the appearance of this new demand, they must now adapt to new conditions. The restoration abandoned traditional cuisine and finally launched an international gastronomic character" Garay & Cánoves (2011, p. 660).

'Gastronomic tourism refers to that branch of the sector where persons make trips to destinations where the local food and beverages are the main motivating factors for travel' (Skift, 2017). Many destinations like Jamaica, Israel, Cayman Island, Australia, etc., are putting local chefs and local food at the center of their marketing campaigns (Skift, 2017). In fine, the objective is to bring visitors closer to local and help them better understand life in the country. Research show that 80% of culinary travelers participate in non-restaurant, food-related activities (visiting winery, eating with local family, or taking cooking classes with local chefs) while on vacation (Skift, 2017). Indeed, Choe, Kim, & Cho (2017); and Therkelsen (2016) explain that local food contributes to: visitors' experience of regional culture; give them a sense of the place, while increasing the earnings of local food producers and tourism business alike; increase social bonds locals/tourists. This vision is also shared by Hjalager & Johansen (2013) who highlighted a 'new rural paradigm' in Denmark where farmers are involved in small-scale food tourism orientated entrepreneurship. More importantly, Hjalager & Johansen (2013, p. 417) explained that this strategy is benefiting 'the wellbeing of both tourists and food producers.'

There are more than 70 definitions of 'sustainable development' (Sharpley, 2000). Destinations now consider sustainability as a competitive advantage to attract visitors, hence the growth of ecotourism, the fastest-growing tourism sector globally. Indeed, consumers are ready to pay more to spend their holidays in destinations considered as sustainable. That said, much of the tourism industry so far has failed to be sustainable hence the reason why Sharpley (2000, p. 1) claimed that 'sustainable development cannot be transposed onto the specific context of tourism.' In this line of thought, Visser (2015) explains that sustainability can be achieved through: Deep reforms; technology innovation; and stakeholders' engagement. Based on the above, gastronomy appears as a sustainable form of tourism as it involves a variety of stakeholders that are all benefiting from the activity. Apps have also been developed by some DMOs to promote destination's cuisine and food culture (Choe et al., 2017). Gastronomy tourism also addresses many issues that destination marketing organizations (DMOs) and locals have recently voiced.

Séraphin, Butcher, & Korstanje (2016) explained that it is very important to educate tourists at pre-visit stage because they very often unfairly judge some destinations. Séraphin, Butcher, & Korstanje (2016) also explained that a good pre-visit stage educational tool must be able to provide: (i) potential visitors experience of the destination before the physical travel; and (ii) a better knowledge of the destination. As a result of these, potential tourists will be turned into active learners and will develop a cognitive, rational, and affective approach of the destination. As for Pilato, Séraphin, & Bellia, & Caescu (2017), they explained that food and food events could be used an educational tool at pre-visit stage (particularly in the case of destination with a negative image). The results of the above academic research are complementary to Gordon's (2017) view as he claimed that gastronomy has the potential to mitigate the negative impacts of mass tourism at visitation stage.

9.2 THEORETICAL DISCUSSION AND IMPLICATIONS

Michel (2000) explains that it is easier for tourists to visit a destination than a destination to welcome tourists as the reasons for visiting are very clear, whereas the benefits of welcoming visitors are not always well defined and managed. In this line of thought, Séraphin (2014); and Korstanje

(2012) explained that the development of the tourism industry can impact negatively on the authenticity of some destinations. As an example, they explain that the name of some places or types of food could be adapted to suit visitors. On that basis, local protesters could for instance argue that tourists are impacting negatively on the quality of food provided by some restaurants which are now aiming for quantity instead of quality. Protesters could even go a step further by banning tourists from cultural events like local food festivals on the basis that cultural events contribute to maintain a link between the members of a group (Booth, 2016) and if overexploited for commercial purpose could be denaturized (Séraphin & Nolan, 2014). This could trigger the emergence of local slow food organizations and/or vintage food activists aiming for a return to authenticity and patriotism. That said, such a radical approach could impact negatively on the image of the destination. This is all the more a problem as image is a key factor of appeal for a destination (Gartner, 2000).

9.3 TOURISM: A JANUSIAN ACTIVITY

For Sanchez & Adams (2008, p. 27) the tourism industry is a very complicated industry because of its 'Janus-faced character.' They explained that while tourism can address some issues like hard currency, it comes with issues like social inequality, racism, prostitution, corruption. For Sanchez & Adams (2008, p. 43) a 'potential can arise in the uneasy marriage between tourism and nation-building.' From this quote, it is legitimate to wonder if tourism could ever be a sustainable activity. Our case, 10 years later after Sanchez & Adams' (2008, p. 27) research, reminds us that the nature of tourism has not changed. Current issues in tourism (more particularly in Europe) indicate that the development of tourism has led to the development of anti-tourism movements. The development of the industry is defeating its objectives. Because of the Janus-faced character of tourism, it seems that an ambidextrous management of the industry is required. Indeed, Smith (2017); and Mihalache & Mihalache (2016) explain that organizational ambidexterity is all about embracing two opposites: *exploitation* (focusing on existing customers and/or markets) and *exploration* (creative and innovative aspects). This approach is presented by Smith (2017) as important for the survival of any organization on both short-term and long term. Gastronomy tourism presents the advantage of being: (i) exploitative as it relies and uses local products (therefore provides incomes for local producers and sheds lights on their expertise that might be unique); and (ii) explorative as it gives a new sustainable option to DMOs in terms of promotion of destinations.

KEYWORDS

- **ambidextrous marketing**
- **consumption**
- **COVID-19**
- **food tourism**
- **globalization**

REFERENCES

Adams, P. D., & Parmenter, B. R., (1995). An applied general equilibrium analysis of the economic effects of tourism in a quite small, quite open economy. *Applied Economics, 27*(10), 985–994.

Ashworth, G. J., (2009). Do tourists destroy the heritage they have come to experience? *Tourism Recreation Research, 34*(1), 79–83.

Becker, D., (2006). *The Essential Legal Guide to Events*. Cape Town: Dynamic Publishing.

Booth, A., (2016). Negotiating diasporic culture: Festival collaborations and production networks. *International Journal of Event and Festival Management, 7*(2), 100–116.

Bressanello, A., (2017). *Some European Cities May Begin Limiting Tourism After Angry Backlash,* Huffingtonpost.com. Available from: http://www.huffingtonpost.com/entry/europe-tourism-backlash_us_59885258e4b041356ec0e5ad?ncid=engmodushpmg00000003 (accessed on 20 January 2022).

Buckley, J., (2017). *Florence Launches Campaign Telling Tourists How to Behave*. Retrieved from: https://www.cntraveller.in/story/florence-launches-campaign-telling-tourists-how-to-behave/ (accessed on 20 January 2022).

Butler, R. W., (1980). The concept of a tourist area cycle of evolution: Implications for management of resources. *The Canadian Geographer, 24*(1), 5–12.

Choe, J. Y., Kim, J. H., & Cho, M. S., (2017). A comparison of food contents used by official tourism organizations' mobile applications. *Journal of Gastronomy and Tourism, 2*, 203–216.

Cohen, E., & Avieli, N., (2004). Food in tourism: Attraction and impediment. *Annals of Tourism Research, 31*(4), 755–778.

Cohen, E., (1972). Toward a sociology of international tourism. *Social Research*, 164–182.

Coldwell, W., (2017). *First Venice and Barcelona: Now Anti-Tourism Marches Spread Across Europe*. Retrieved from: http://www.theguardian.com (accessed on 20 January 2022).

Curtin, S., Richards, S., & Westcott, S., (2009). Tourism and grey seals in South Devon: Management strategies, voluntary controls and tourists' perceptions of disturbance. *Current Issues in Tourism, 12*(1), 59–81.

Dong, E., Morais, D., & Dowler, L., (2003). Ethnic tourism development in Yunnan, China: Revisiting butler's tourist area lifecycle. In: *Proceedings of the Northeastern Recreation Research Symposium*. Bolton Landing, NY.

Fields, K., (2002). Demand for the gastronomy tourism product: Motivational factors. *Tourism and Gastronomy*, 36–50.

Fischler, C., (1988). Food, self and identity. *Information (International Social Science Council), 27*(2), 275–292.

Garay, L., & Cánoves, G., (2011). Life cycles, stages and tourism history: The Catalonia (Spain) experience. *Annals of Tourism Research, 38*(2), 651–671.

Gartner, W. C., (2000). Image. In: Cooper, C., & Hall, M., (eds.), *Contemporary Tourism: An International Approach* (pp. 223–229). London: Elsevier.

Gay, J. C., (2012). Why is tourism doing poorly in overseas France? *Annals of Tourism Research, 39*(3), 1634–1652.

Gordon, Y., (2017). *Lisbon's 'We Hate Tourism' Tours Teach Visitors How to Travel Responsibly*. Independent.co.uk Available from: http://www.independent.co.uk/travel/europe/lisbon-tourists-we-hate-tourism-tours-locals-responsible-travel-portugal-a7919571.html (accessed on 20 January 2022).

Hall, C. M., & Mitchell, R., (1998). *We are What We Eat: Food, Tourism and Globalization.'* paper presented at innovative approaches to culture and tourism, ATLAS Conference. Rethymno, Crete.

Hall, C. M., & Sharples, L., (2004). The consumption of experiences or the experience of consumption? Introduction to the tourism of taste. *Food Tourism Around the World*, 1.

Hammond, M., & Wellington, J., (2013). *Research Methods: The Key Concepts*. London & New York: Routledge.

Hewlett, D., Harding, L., Munro, T., Terradillos, A., & Wilkinson, K., (2016). Broadly engaging with tranquility in protected landscapes: A matter of perspective identified in GIS. *Landscape and Urban Planning, 158*, 185–201.

Hjalager, A. M., & Johansen, P. H., (2013). Food tourism protected areas – sustainability for producers, the environment and tourism? *Journal of Sustainable Tourism, 21*(3), 417–433.

Huffingtonpost, (2017). *Some European Cities Many Begin Limiting Tourism After Angry Backlash*. Retrieved from: http://www.hufffingtonpost.com (accessed on 20 January 2022).

Jeynes, J., (2002). *Risk Management: 10 Principles*. Oxford: BH.

Kettle, M., (2017). *Mass Tourism is at a Tipping Point – But We Are All Part of the Problem*. theguardian.com, Available from: https://www.theguardian.com/commentisfree/2017/aug/11/tourism-tipping-point-travel-less-damage-destruction (accessed on 20 January 2022).

Kivela, J., & Crotts, J. C., (2006). Tourism and gastronomy: Gastronomy's influence on how tourists experience a destination. *Journal of Hospitality & Tourism Research, 30*(3), 354–377.

Korstanje, M., (2012). The influence of language in the travel agencies name: The case of buenos Aires. *The Journal of Hospitality and Tourism, 1*(10), 127–134.

Kruczek, Z., & Szromek, A., (2011). Using the butler cycle for interpreting tourist attractions, based on the example of the salt mine in Wieliczka. *Folia Turistica, Special Edition "The Master Classes," Part II: From the Work of Polish Tourism Researchers, 25*, 2.

Leadbeater, C., (2017). *Anti-Tourism Protesters in Barcelona Slash Tyres on Sightseeing Buses and Rental Bikes*. Available from: http://www.telegraph.co.uk (accessed on 20 January 2022).

McCool, S. F., & Martin, S. R., (1994). Community attachment and attitudes toward tourism development. *Journal of Travel Research, 32*(3), 29–34.

McDowall, S., & Choi, Y., (2010). A comparative analysis of Thailand residents' perception of tourism's impacts. *Journal of Quality Assurance in Hospitality & Tourism, 11*(1), 36–55.

Michel, F., (2000). Hosts and others. Tourism and otherness. *Spaces,* 171.

Mihalache, M., & Mihalache, O. R., (2016). Organizational ambidexterity and sustained performance in the tourism industry. *Annals of Tourism Research, 56*, 128–163.

Orams, M. B., (1996). A conceptual model of tourist-wildlife interaction: The case for education as a management strategy. *Australian Geographer, 27*(1), 39–51.

Paris, N., (2017). *Tourists Have Turned Oxford Into 'Hell' Local Claim*. telegraph.co.uk, Available from: http://www.telegraph.co.uk/travel/news/oxford-tourist-hell-overcrowding-residents-locals-complain/ (accessed on 20 January 2022).

Parkes, C., Scully, J., & Anson, S., (2010). CSR and the 'undeserving': A role for the state, civil society and business. *International Journal of Sociology and Social Policy, 30*(11, 12), 697–708.

Petkar, S., (2017). *'Spain is Quite Full' Tourists Urge to Travel Elsewhere as Protests Grip Holiday Hotspots*. express.co.uk, Available from: http://www.express.co.uk/news/world/840042/barcelona-spain-holidays-anti-tourist-protests-go-home-graffiti-violence-tui-travel (accessed on 20 January 2022).

Pilato, M., Bellia, C., & Sèraphin, H., (2016). Street food and food safety: A driver for tourism? *PEEC 2016 International Conference, Bucharest University of Economic Studies*. Bucharest (Romania).

Pilato, M., Sèraphin, H., Bellia, C., & Caescu, S. C., (2017). Challenging the negative image of destinations at pre-visit stage using food and food events as an educational tool: The case of Romania. *ETIMM Conference*. (Romania).

Plog, S. C., (1974). Why destination areas rise and fall in popularity. *Cornell Hospitality Quarterly*.

Powell, C., (2013). *How to Deliver Outstanding Corporate Events*. London: Lulu Publishing.

Puczko, L., & Ratz, T., (2000). Tourist and resident perceptions of the physical impacts of tourism at Lake Balaton, Hungary: Issues for sustainable tourism management. *Journal of Sustainable Tourism, 8*(6), 458–478.

Reynolds, P. C., (1993). Food and tourism: Towards an understanding of sustainable culture. *Journal of Sustainable Tourism, 1*(1), 48–54.
Richards, G., (2002). Gastronomy: An essential ingredient in tourism production and consumption. *Tourism and Gastronomy, 11*, 2–20.
Ritzer, G., (1993). *The McDonaldization of Society*. Thousand Oaks: Pine Forge Press.
Rutherford, S. J., (2008). *Risk Management for Meetings and Events*. Abingdon: Routledge.
Sanchez, P. M., & Adams, K. M., (2008). The Janus-faced character of tourism in Cuba. *Annals of Tourism Research, 35*(1), 27–46.
Séraphin, H., & Nolan, E., (2014). Voodoo in Haiti: A religious ceremony at the service of the 'houngan' called 'tourism.' In: Frost, W., & Laing, J., (eds.), *Rituals and Traditional Events in the Modern World*. New-York: Routledge.
Séraphin, H. (2014) Les jeux d'influences dans le tourisme: Cas d'Haïti, Journal *of Haitian Studies, 20*(2), 66–86.
Séraphin, H., Ambaye, M., Gowreesunkar, G. B., & Bonnardel, V., (2016). A marketing research tool for destination management organizations' logo design. *Journal of Business Research, 69*(11), 5022–5027.
Séraphin, H., Butcher, J., & Konstanje, M., (2016). Challenging the negative images of Haiti at a pre-visit stage using visual online learning materials. *Journal of Policy Research in Tourism, Leisure and Events*. doi: 10.1080/19407963.2016.1261146.
Séraphin, H., Pilato, M., & Bellia, C., (2017). Entrepreneurship in the tourism sector: A risky activity in French overseas departments? *International Conference: The Risk in Contemporary Economy*. University of Galati, Romania.
Sharpley, R., (2000). Tourism and sustainable development: Exploring the theoretical divide. *Journal of Sustainable Tourism, 8*(1), 1–19.
Sissic, J., (2017). *Should We Worry About the Development of Anti-Tourism?* TourMag.com, Available from: http://www.tourmag.com/Faut-il-s-inquieter-du-developpement-de-l-anti-tourisme_a83390.html (accessed on 20 January 2022).
Sissic, J., (2017). *Tourismophobia: How to Fight Against the Irresponsibility of the Tourist and his Excesses*? [Tourismophobie: comment lutter contre l'irresponsabilité du touriste et de ses excès?] TourMag.com, Available from: http://www.tourmag.com/Tourismophobie-comment-lutter-contre-l-irresponsabilite-du-touriste-et-de-ses-exces_a88607.html (accessed on 20 January 2022).
Skift, D. P., (2017). *Israel Looks to Food Tourism to Attract a Different Kind of Visitor*. Retrieved from: https://skift.com/2017/07/26/israel-looks-to-food-tourism-to-attract-a-different-kind-of-visitor/ (accessed on 20 January 2022).
Smith, S., (2017). Organizational ambidexterity: Welcome to paradox city. *Human Resources Management International Digest, 25*(1), 1–3.
Tapper, J., (2017). *As Touting for Punt Trips Becomes a Crime, is Tourism Overwhelming Britain's Cities*. Retrieved from: http://www.theguardian.com (accessed on 20 January 2022).
Telfer, D., & Wall, G., (1996). Linkages between tourism and food production. *Annals of Tourism Research, 23*(3), 635–653.
Therkelsen, A., (2016). Taking stock of the new Nordic cuisine at Danish coastal destinations. *Journal of Gastronomy and Tourism, 2*, 15–25.

Visser, W., (2015). *Sustainable Frontiers: Unlocking Change Through Business, Leadership and Innovation*. Sheffield: Greenleaf Publishing.

Yazdi, S. K., & Khanalizadeh, B., (2017). Tourism demand: A panel data approach. *Current Issues in Tourism, 20*(8), 787–800.

CHAPTER 10

Leisure Inequalities in South America: Argentina as a Main Study Case in a Post-COVID Scenario

MAXIMILIANO E. KORSTANJE

Department of Economics, University of Palermo, Larrea ST 1079, Buenos Aires, Argentina

ABSTRACT

Without any doubt, leisure studies have been experienced notable growth over the recent decades. Leisure has been captivated scholars and sociologists globally. From its onset, sociology overlooked leisure as a serious object of study. Since Durkheim, sociology was not interested in studying leisure as a valuable instrument of society. From the Roman to British empires, leisure not only revitalizes social frustration but also keeps society united. In the post-COVID-19 days, we discuss the emergence of the new forms of leisure practices as well as tourism behavior worldwide. With a focus on Argentina as the main study case, the present chapter holds the thesis that COVID-19 revived not only long-dormant discourses but also inter-class asymmetries and inequalities in the Global South.

10.1 INTRODUCTION

Leisure studies have advanced a lot in recent decades, and of course taking different definitions. No less true seems to be that there are some academic disputes revolving around its etymological nature. The term

Post-Disaster and Post-Conflict Tourism (An Updated 2nd Edition), Maximiliano E. Korstanje, Hugues Seraphin, and Vanessa G. B. Gowreesunkar (Eds.)

leisure comes from French '*loisir*' which means free time. Ancient Greeks used the term *Scholé* (σχολή) to denote leisure but in the time its meaning has substantially changed. Ancient leisure practices varied on modern ones. Historians acknowledge that *Scholé* was a social institution reserved for philosophical contemplation not strictly linked to entertainment like in the modern culture. In passing the time, the term '*Scholé*' mutates to Schola where come from the modern terms school and scholar (Eng.). This suggests that leisure and education are inevitably entwined (Balme, 1984). During the days of the Roman Empire, Schola sets the pace to a new term, '*otium*' (leisure) which was only invoked to refer to evasion. As strictly opposed to neg-otium, the term signals the right of entertainment each free citizen enjoys. Having said this, leisure accompanied not only the Roman life but also the history of sedentary societies without mentioning modern empires. Leisure unites frustrated citizens sublimating their fears in a unique experience into the same spirit, but at the same time, it separates strangers (Kyle, 2012). Over centuries, leisure played a crucial role not only in legitimating political authorities but also in keeping society united. Countless social frustrations, which are proper of the inconsistencies of society, are finally sublimated and revitalized in the leisure space. Hence, leisure should be seen as culturally enrooted in politics (Korstanje, 2009). Whenever a state of crisis takes a hit, no less true is that leisure activities show serious inter-class imbalances which are previously covered. This seems to be exactly what happens with the global pandemic originated by the SARS-CoV2 virus (known as COVID-19) that paralyzed the entire globe. Countries adopted restrictive measures to stop the pandemic affecting seriously the tourism industry. The pandemic brings globalization to a standstill causing serious economic losses for the industry. At the same time, the material asymmetries between rich and poor countries became evident. A whole portion of COVID-19 vaccines was finally monopolized by developed economies. In this grim context, thousands of Argentinean tourists fled to the US in quest of vaccination. At the end of June, President Fernandez disposed of a drastic reduction of the quota of tourists who can return to the country. The quota passed from 2,000 to 600 authorized tourists to come back (estimating only 3 commercial flights per day). This ad-hoc disposition generated not only a humanitarian crisis but also many tourists who were stranded without financial assistance. This research gives the preliminary outcomes of some interviewees conducted over tourists stranded in Miami city. Based on the need of understanding their fears, expectances, and feelings, the present work discusses not only the philosophical dilemmas of vaccine tourism but also the economic inequalities between the Global North and South.

In his seminal text, '*Tourist and Vagabonds: Heroes and Victims of Postmodernity*,' Zygmunt Bauman (1996) calls the attention to the opposition between vagabonds – an underclass formed by migrants and asylum seekers—and global tourists. Per Bauman, vagabonds, and tourists move but with different degrees of autonomy – or dependency. Post-modernity re-constructs a hierarchal class organization marked by the freedom each citizen has to choose his own itinerary. Bauman distinguishes the archetype of pilgrims who delay their gratification from tourists who often motivate by a strategy of pleasure-maximization. The strategy of the pilgrim is certainly based on the needs of salving the future. Rather, the tourist eradicates the present stranded in a timeless itinerary. The figure of the tourist is not in quest of an identity, but the simple avoidance of being fixed. The tourist never belongs to the visited place but what is more important, the tourist moves to be closer but at the same time he keeps distance from the locals. Inside this symbolic bubble he feels finally safe. Having said this, locals never interrogate tourists because they have the power to exercise their right to travel. Basically, the tourist travel engages with the act of being gratified by the feeling of being in control interacting occasionally with "others" while switching on or off that connection to my discretion. As a result of this, each (tourist) experience replaces cyclically an older one in an inconsequential logic. The decision to travel or leaving home is precisely not to be sick at home (home-sick). In this way, the tourist becomes a real wanderer because he travels with no bounds. Unlike the pilgrim who is suspected, the tourist is never abandoned but protected from the potential dangers during his trip. We live in a postmodern world that is inhospitable to pilgrims, which means to those who are identity-builders. They struggle to make a solid world while modernity goes towards a liquid state. The control of time gives not only more gratification but also further authority over others. As he puts it, the more freedom of choice the citizen has the higher status in postmodern society. This suggests that global capitalism creates material asymmetries strictly based on the autonomy to travel. This begs a couple of pungent questions: how do social scientists apply this concept on a post-COVID-19 world? What does happen with tourists in a world where they are the undesired guests?

10.2 THE NATURE OF LEISURE: A SHORT INTRODUCTION

The act of conceptualizing or defining leisure seems to be a hard task to accomplish. There are a lot of discrepancies as well as academic schools which have adopted their definition, while in other cases a multiplication of synonymous terms surfaced (Stebbins, 2011). Given this context, sociologists, and anthropologists have systematically overlooked the role of leisure to keep the society united (Wilson, 1980; Dunning & Rojek, 1992). As a young Norbert Elias recognized, leisure was not on the agenda of the founding parents of sociology, nor in the spirit of their followers (Elias & Dunning, 1986). From its outset, sociology focused on social maladies which include criminality, anomie, and the acculturation process as the main object of study relegating leisure as a naïve (alienable) practice (Korstanje & George, 2021). No matter the time or culture, leisure accompanied the history of sedentary tribes as well as the consolidation and expansion of empires. At a closer look, the political authority depends upon the success in articulating leisure rites; Roman emperors disposed of Gladiators' fights, public banquets, or baths, to enhance their legitimacy. Leisure amalgamated roman citizens in a climate of brotherhood regardless of their status or class, but at the same time, it relegated to those who are not considered free subjects. This happened simply because leisure inscribes into the logic of politics (Korstanje, 2009). In 1899, American economist Thorstein Veblen published his seminal book the "*Theory of Leisure Class.*" He acknowledges that society divides into productive and unproductive classes. While the former is drawn by technicians and workers, the latter reserves the monopoly of the pecuniary emulation. Pecuniary emulation transfers status to the privileged class which manufactures the surplus excess. This leisure class is formed by politicians, scholars, priests, and warriors (soldiers). They control but at the same time, they are exempted from the productive system. To put the same simply, leisure class is naturally unproductive and adopts a parasitic lifestyle (Weblen, 2017). Having said this, some voices devoted considerable time and effort trying to decipher the nature and evolution of modern leisure practices. Spanish theorist Frederick Munné is strongly moved to resolve the theoretical fragmentation revolving around leisure. Although his texts are not translated to English, he has shed light on the cultural studies about the nature of leisure. He starts from the premise that leisure should be understood as a political (intermixed) act which guides the subject to philosophical contemplation. Any free citizen debates between two opposing forces. On one hand, leisure represents the structuration of rules and habits which makes life something predictable and secure. On another, the human needs to escape – at least temporarily – from standardization and mechanization. While combining both tendencies, leisure engages with a much deeper revitalization process which inscribes in the symbolic core of society (sentiment of compensation) (Munné, 1980; Munné & Codina, 2002). Let's remind readers that Munné's legacy is mainly inspired by two authoritative voices in leisure studies: Erich Weber and David Riesman. Both – from different angles—give different conceptual models to understand leisure as a new modern sociological phenomenon. David Riesman develops a typology based on the connection of the man with a specific sociological character. Societies gradually evolved from inner-directed character to other-directed type, where the self is previously moved by the peer's esteem. The inner-directed type is based on the puritan culture where the self is individually cultivated. The industrial process installs a curiosity for the "Other" transforming radically the self. Travels and expeditions, which are inspired by a quest for new sites or unique experiences, are particularly a quintessential feature of other-directed characters. In a nutshell, travels, discovery, and tourism are possible in other-directed societies which remain overt to new experiences. The interest for the "Other" crystallizes in the needs of economic expansion, which was proper of capitalism. Therefore, colonial travels and colonialism are historically inter-connected (Riesman, Glazer, & Denney, 2001). Weber's legacy is particularly very interesting as well. For Weber, leisure is described by a continuum of regeneration, compensation, and idealization. Whereas regeneration appeals to the physical need of evasion and relaxation, compensation is feasible only under the conditions where the idealized goals are reached. The idealization mainly marks a contemplation process that takes the lead when a person meets a state of self-fulfillment (Weber, 1969; Rojek, 2013). At a closer look, leisure not only takes different forms and meanings which vary by time and society, but paradoxically it serves as an ideological instrument of domestication (following carrot-and-stick methods). Leisure alternates significant and complex functions of revitalization (change) and preservation which merit being investigated (Trunfio, 1991). It is safe to say this concern inspired sociologists Norbert Elias & Erich Dunning (2008) in their work Quest for Excitement: sport and leisure in the civilizing process. Per their viewpoint, leisure practices are originated in ancient Greece, particularly evolving according to what they dubbed as the civilization formation. Leisure appears to be inextricably intertwined with the evolution of modern capitalism. As a mitigated, if not controlled version of wars, leisure ignites a process of rivalry and cooperation to regulate internal violence (or inter-group conflicts). Elias & Dunning focus their study on hooliganism explaining that modern capitalism represses and re-channel human emotions towards the private sphere. Public events or sports games activate a bearable state of rivalry originally articulated to domesticate violence. In consonance with Weber, Elias, & Dunning (2008) hold the thesis that the civilizatory process engages with pleasure and joy but opening the doors to risky situations. Defined as a disciplined form of war, leisure interrogates furtherly the philosophical existence of self, leaving it from the humdrum routine, but exposing it to violence. Some philosophers accept leisure is based on a manifest paradox. As Karl Spracklen puts it, leisure practices associate with the question of freedom, the free choice that any citizen enjoys. Nonetheless of this fact, the bourgeoisie liberty cannot be divorced from the local law, or in his terms, the rules of the city. In sum, theorists debated according to the three-clear axes: leisure as a free choice, leisure as socially constrained, and leisure as a completely free choice. While liberals adopt leisure as the sign of free democratic practice, which consolidates democratic institutions, structuralism stresses the role of structure which is finely ingrained in the productive system. Postmodern writers allude to leisure as a complex net of relations inscribed in politics; however, this interplay between liberty and constraints seems to be far from being a contradiction as the literature punctuates. The rites of leisure provide citizens with a temporal free choice to play, and in so doing, they make the decision to internalize the rules of society; this frequently occurs because the authority of chiefdom emanates how successful leisure practices are regulated. In essence, leisure should be studied as an inherent part of politics (Spracklen, 2011, 2013).

Last but not least, Chris Rojek debates hotly to what extent the agent is free or not to practice leisure. He coins the embodiment and emplacement of the term. While embodiment signals to the biological nature of the body, which is all days dying, the emplacement determines where the agent moves. Leisure, in this way, activates human reciprocity enhancing a deteriorated sense of us-living-together placating the devastating consequences of inter-class conflict (in Hobbesian terms avoiding the war of all against all) (Rojek, 2005a, b, 2013). The section has shortly introduced readers to the discussions that revolve around the nature of leisure. The next one delves into the evolution of modern leisure practices (with a focus on modern tourism) in a post-COVID-19 world.

10.3 LEISURE AND TOURISM AFTER COVID-19

From its outset as a discipline, tourism research has widely incorporated and adopted profit-based definition of tourism nature. We have discussed the nature of leisure and of course, agree with the argument tourism is more than a mere industry (Salazar & Graburn, 2014). Within the constellations of tourism sociology, the discussion seems to be far to be closed (Craik, 2002; Theobald, 2005; Tribe, 2010). For American anthropologist, Dean MacCannell tourism should be recognized as a powerful vehicle that avoids social fragmentation. In any tribal organization, as MacCannell said, the totem, as well as religion, endorses the authority from the figure of the founding parents to chiefdoms. The totem is the only source of political authority in the tribe. In industrial societies, the secularization process displaces religion to a peripheral position leaving a gap which is filled by tourism. In other terms, tourism keeps the inevitable social fragmentation provoked by global capitalism. Tourists look for the essence of authenticity laying a paradoxical condition which associates to the imposition of fabricated forms of relations he named as "staged authenticity." Based on a Marxist tradition, MacCannell holds that the structure of tourism consumption alienates the modern workforce through the absorption of the amassed surplus. After all, tourism globally accelerates proper economic inequalities among rich and poor nations (MacCannell, 1973, 1976, 2002). In opposition to MacCannell, John Urry approaches tourism as a global force resulted from two combined factors. On one hand, he speaks to us of the technological breakthrough which leads capitalism towards a rapid acceleration of travel timeframes. In consequence, global capitalism expands according to the basis of a decentralized form of production (economies of signs). To put this in straights, capitalism engenders high-mobility societies. On another, this cultural matrix creates a type of different tourist gazes which marks what landscapes can be seen or not. Here there is a type of desirability that ultimately captivates his attention. Like MacCannell, Urry recognizes the paradoxical condition where tourism operates. In one direction, tourism cultivates democratic and prosperous societies, but on the contrary, it enlarges the gap between the have and have-nots (Urry, 1990, 2002; Lash & Urry, 1993). Both agree that tourism is a modern activity widely designed and molded by the industrial revolution. Most certainly, Erik Cohen goes on to a mix-balanced argumentation that deserves some discussion. Taking some distance from MacCannell and Urry, he argues convincingly that tourism should not be contemplated within the figure of escapement only. Rather, neither the estheticized modernity nor industrialization suffices to explain its nature. Cohen defines tourism as a type of commercialized hospitality that creates liminoid spaces of play, where rules are temporarily subverted. In Cohen, the notion of play is vital to frame conceptually tourism (Cohen, 1972, 1979, 1988). In the present research, we go beyond. Tourism should be defined as a rite of passage, so to speak as a sub-type of leisure practice. Anthropologically speaking, any rite of passage recreates the condition for escapement while accommodating the self into the new in-group rules. The sense of renovation and social status is of paramount importance to understand rites of passage. Baptism, holidays, weddings, and public events are considered rites of passage where human groups (society) discuses and renovates the legitimacy of their authorities. In tourism practices, citizens play to be what they dream to be, but they are not in reality. This liminoid space well-described by Cohen is based on what we dubbed as a "dream-like factor." Like dreaming, tourism revitalizes all our frustrations into a renovated – and integrated-self. Tourism traverses the industrial process. Ancient Romans had a term '*feriae*,' which was a lease given to citizens for visiting relatives for three months. Likewise, Feriae – as well as the passenger exchange – helps Roman authorities not only keep the provinces symbolically embroiled to Rome, while enhancing reciprocity, but they assured a rapid military presence in case of sedition. For some reason which is hard to precise here, classic historians overlooked this point turning their attention on the industrial revolution as the historical moment that marks the rise of tourism. The medical epoch was characterized by low-mobilities as well as a great political fragmentation. However, ancient history gives much evidence about leisure travels which may be very well equated to modern holidays. Today, we must confess, in modern languages like German (die Ferien) or Portuguese (das Ferias), the etymological influence of this ancient (roman) institution appears to be unquestionable (Korstanje, 2018b; Korstanje & George, 2022). Not surprisingly, similar leisure travels may be traced back to ancient empires such as Babylonians, Assyrians, Egyptians, and Nubians (only to name a few). It is safe to say that tourism can be defined as "the maiden of empires" (Korstanje, 2018a; Korstanje & George, 2022). What is equally important, like leisure, tourism changes according to the structural transformations of the society (Rowe, 2002; Krippendorf, 2010; Elkington, 2015). The global pandemic, whose first cases were originally reported in Wuhan, China, has brought devastating and long-lasting effects for the tourism industry as well as for travel behavior. Governments unilaterally adopted restrictive steps and measures which changed day-to-day life. These measures included the closure of borders and airspaces, as well as the imposition of strictest lockdowns and social distancing (Gossling, Scott, & Hall, 2020; Prideaux, Thompson, & Pabel, 2020). The figure of the "non-Western Other" who historically was considered as an object of study or curiosity for European ethnographers and colonial voyage are now suspects to disseminate a lethal virus. Tourists are no longer ambassadors of democratic nations, now they are treated as terrorists (carriers) who knowing or not infects "other vulnerable citizens" (for further details see Korstanje, 2020, 2021). As Korstanje & George (2021) lamented, tourists, have become "undesired guests." Some studies recently document long-dormant prejudices or chauvinist discourses against Asian tourists or new geopolitical tensions among nation-states. The borders of global geography have been radically altered to a more fractured (if not feudalized) world (Korstanje & George, 2021, 2022). This raises a moot point: is COVID-19 the opportunity of a new re-start for the tourism industry or simply a sign of exhaustion that exhibits – of course affirming – the material asymmetries between the Global North and South?

As the previous argument is given, Higgins-Desbiolles, Doering & Bigby depart from the premise that the COVID-19 opens the doors to a new opportunity for re-socializing tourism into a new type of consumption that leads towards ecological justice. Per their stance, tourism is a social force in nature that needs to be socialized to balance profits and human action. In this way, governments will somehow reach a durable state of political stability and prosperity. Market-centered tourism should set the pace to community-centered tourism, as these experts adhere. Although the effects of COVID-19 on the tourism industry seem to be devastating and long-lasting, as Raoul Bianchi (2022) adheres, there are few signs of a much deeper renewal or restart. The concentration of power, wealth, and profits in the hands of the global north has created considerable financial asymmetries among nations. Corporate-managed extractive institutions have successfully monopolized not only knowledge production but also capital distribution globally. COVID-19 reveals some hidden injustices while putting limits to untrammeled growth, but it is far from reversing them. It is safe to say that the pandemic has accelerated and brought countless socio-economic effects in the tourism industry worldwide. Nowadays not only those countries dependent on tourism are seriously affected but also the future of the industry remains uncertain (Romagosa, 2020; Sigala, 2020; Rogerson & Rogerson, 2020). A cross-analysis study conducted by Farzanegan et al. (2021) brings three important aspects into the foreground. At a first glimpse, those nations highly exposed to international flights – because of their dependence on international tourism – registered further deaths caused by COVID-19. Secondly, COVID-19 shows affecting further over-crowded urban than rural destinations. Third, the international tourism flows correlate directly to the virus expansion. This reveals that the tourism industry is the main carrier and victim of COVID-19 (Baum & Hai, 2020). The real impact of COVID in the industry is manifold and complex. Its long-lasting socio-economic effects include political instability, unemployment, and economic downturn, discourses associated with separatism, inter-class conflicts, geopolitical tensions, or chauvinist (racist) narratives without mentioning hostility against Asian tourists (Mostafanezhad, Cheer, & Sin, 2020; Korstanje & George, 2021; Rose-Redwood et al., 2020). These consequences become more acute in small and dependent tourism destinations (Carr, 2020). What seems to be more important, the global economy sets the pace to a new (feudalised) fractured world widely atomized where current travel behavior has been radically shifted. New emerging travel bans, or heavy migration requirements are asked to foreigner visitors, accompanied by new travel regulations such as PCR tests, health insurances or health passports. The "Other" has become "an undesired guest" who is suspected to be a carrier of a lethal virus. COVID-19, so to say, affirms the already-current crisis of hospitality in the West (Korstanje, 2017; Korstanje & George, 2021, 2022; Barbosa et al., 2021; Ritzer, 2007; Yar & Tzanelli, 2019). In consonance with this, Tzanelli & Yar (2020) eloquently describe the climate of anti-hospitality in the Western capitalist marketplace. As they note, the tourism industry is facing an atmosphere of inhospitality as never before. New practices where the tourists pay to be humiliated, kidnapped or even tortured have recently come to stay. They hotly debate Barber's original thesis to make terror a consumable spectacle. The "Other's death" has become the ultimate form of pleasure maximization (Korstanje, 2016; Tzanelli, 2016). Lastly, the monopolization of vaccines in the hands of rich nations still is an issue which evinces the asymmetries between the global North and South. While rich countries monopolize a whole portion of vaccines, fewer resources countries are left to cope with the pandemic in unfavorable conditions. As a result of this, poor nation citizens deprived of the vaccine made accommodation in Europe and the US. Of course, Argentina seems not to be an exception. The lack of vaccines associated with the economic crisis the country goes through pushes thousands of Argentinean tourists to fly to the US in quest of better opportunities to be inoculated. These tourists, who are high purchasing consumers, are psychologically motivated by countless goals. President Alberto Fernandez and his ministers overtly rivaled and condemned these tourists who flying abroad contributes to the virus dissemination. They were demonized as "the first-class tourists" who violated the lockdown putting the rest of compatriots at risk. In the successive days, the government suddenly disposed of the airspace and border closure leaving thousands of tourists stranded abroad. The new tourism niche, which is known as vaccine tourism, evinces one of the inequalities of capitalism as well as the inter-class ideological conflicts that traverse society. For the sake of clarity, the point is addressed with further detail in the next section.

10.4 FLYING TO MIAMI: VACCINE TOURISM

It is important not to lose sight of the fact that the restart of the tourism industry, as well as global trade, depends upon the success of the vaccination process. Many people are inoculated faster, the better for a rapid economic recovery. Health and vaccine tourism share a similar logic which deserves further investigation (Korstanje & George, 2022). In either case, tourists are psychologically moved to receive medical attention deprived at their homelands. This segment is often subject to countless regulations as well as disparities between rich tourist-receiving and poor tourist-delivering societies. For practical domains, population aging, lifestyle changes, and the enhancement of health systems in capitalist economies are key factors that explain health tourism. These tourists are in quest of opportunities to receive health treatment impeded in their countries (Hall, 1992; Garcia Altes, 2005). The case of Cuba seems to be particularly pretty different. As Goordrich (1993) eloquently observes, since the revolution Cuba has historically emphasized the strengths of its health system and devoted resources to situate as a definitive health destination. Over the decades, Cuba opened the doors to a new niche of travel health care services. Health and vaccine tourism correspond with the economic inequalities of health systems across the globe (Smith & Jenner, 2000). Health tourism derives from the decline of the welfare state just after the 70s decade. The globalization process was conducive not only to the dismantling of the welfare state but also to other economic deregulation which facilitated health tourism. Based on the use of digital technologies, cross-border multinationals arrived at under-developing nations to invest in the local health system to ensure standardized health services. Likewise, intra-national medical tourism becomes a fertile segment in the developed and under-developing nations (Reisman, 2010). The opposite is equally true; this standardization leads 1,000 of rich consumers to fly peripheral countries in quest of cheaper health attention and surgery rates. Having said this, transplant tourism is a fertile ground in under-developing countries (Beland & Zarzeczny, 2018). Some voices have alerted on the ethical intersection of human trafficking or organ trafficking with transplant tourism in poor economies. People are pressed to sell their organs or simply are kidnapped as organ harvesters. Medical tourism – like vaccine tourism—evinces the uneven forms of development of nations worldwide (Ambagstsheer, Zaitch, & Weimar, 2013; Budiani-Saberi & Delmonico, 2018; Columb, 2019). Like children abuse, criminal activities associated with organ trafficking are not heavily punished in some countries. Beyond the legality of leisure consumption which appertains to the prosperity of under-developing nations, many criminal organizations are operating in the forms of transplant or sexual tourism; this happens simply because there is a liminoid space where criminality and legality converge (Ryan & Kinder, 1996; Montgomery, 2010). Although the specialized literature in health tourism abounds (Hunter-Jones, 2005; Hall, 2012; Suess, Baloglu, & Busser, 2018), less is said about vaccine tourism in the context of COVID-19. To fill the gap, the present section discusses the ebbs and flows of Argentinean tourists to be inoculated in the US because of the scarcity of vaccines in Argentina as well as the partisan discussion and ideological discrepancies with Alberto Fernandez's administration.

As stated above, the COVID-19 outbreak accelerated a medical crisis in the world stressing the gap between poor and rich countries. The most advanced economies received almost 380 million doses of COVID-19 vaccine while leaving the periphery in a difficult health situation. To the mid of April, Argentina was at a snail pace in the vaccination campaign. Since the US opened the free vaccination for tourists in the mid of March 2021, thousands of tourists fled to the US (preferably Miami where they share the language) to be vaccinated. Not only celebrities and businessmen with higher purchasing power but also desperate mid-class families—whose children had comorbidities—selected Miami as a preferred destination. At the end of June of 2021, Government reduces suddenly the quota of Argentinean tourists who can return to the country – from 2,000 to 600 per day. Without previous notice, this restrictive measure, which was unilaterally adopted, coincides with the worries of health authorities for the Delta strain dissemination. As a result of this, many international flights should be delayed and rescheduled, leaving thousands of tourists stranded abroad. Some of them are deprived of shelter, food, and accommodation. Such a disposition is now upsetting public opinion. In consequence, protests and marches against Fernandez's resolution spread Argentina-wide. The following lines summarize a set of formal and informal interviews conducted on tourists who have been stranded abroad.

10.5 STRANDED ABROAD

The present section contains a small portion of the 10 interviews conducted through Zoom, Google-meet, and Skype to different families stranded in the US and Europe just after Fernandez' polemic resolution. The sample was formed by 30 interviewees (20 males and 10 females) between 25 and 59 years old. The method corresponds with the snow-ball technique, which means each interviewee recommended the next person to consult. The obtaining outcomes are not statistically representative and cannot be extrapolated to other universes.

- ***I Feel Abandoned by My Government***

  As discussed in earlier sections, Fernandez's resolution derived from thousands of tourists stranded in Miami, Florida as well as in different airports worldwide. Marcelo (male, 28 years old) claims:

"I am stranded in the US without financial support and food. I am living with what some friends give me. I feel abandoned by the government, by the state. I have never thought President Fernandez makes a disposition like this. I am thousands of miles away from my home, my family ... fully disappointed of this! If you ask, I do believe the government wants to punish us because we opt for being vaccinated in the US."

This coincides with Marcela (female, 35 years old) who overtly says:

"I am calling to the Aerolineas Argentinas [Argentinean Airlines] since this morning and my flight was scheduled to August 4. I called several times to change that but without any result. I am desperate! I need to come back urgently. Like me, almost 200 compatriots are sleeping everywhere at the airport... this is a clear implicit sanction of this communist government, we live like in Cuba people flies but cannot return."

These interviewees feel abandoned by the state adducing partisan ideologies to castigate those citizens who flew to the US, Miami or First-World destinations. For them, the US is considered a vaccine tourist destination preferred by the easier barriers to get inoculated. The scarcity of vaccines in Argentina pushes many tourists to travel abroad to be vaccinated.

- ***Marginalized Travelers: The Banality of Tourism***

The crisis provoked by the global pandemic divides public opinion n two sides. Those who are in support of the government and the travel bans distinguish themselves from those who claim for their liberties and rights to be respected despite the situation. In the same vein, tourists were trivialized as hedonists and naïve travelers who maximize their pleasure no matter the health condition of the country. On different occasions, President Fernandez and their ministers raged against tourists who violate the lockdown or kept insensitive to the current sanitary context. Juan (male, 59 years old) exclaimed:

"I am annoyed with this government. They treat us as citizens of the second class. The government discriminates against us thinking we are leisure travelers who are part of a privileged elite. I was working in the

US when my work visa expired. Now I need to come back to Argentina. The Government closes the doors to us because they hate us."

Maria (female 22 years old) is stranded in London given the fact that she was a postgraduate student. She replies:

"This pandemic divides the world in two, those who can travel (like us) and those who lost their jobs (because of the failures of the governments to stop the pandemic). This government retaliates on the mobile citizens who can travel abroad. This is more complex than hating it is a new state of the world where tourists are demonized as potential evildoers or carriers of the COVID-19."

In both interviews, participants manifest government undermines them as second-class citizens, marginalizing them into a peripheral position. Lastly, Government is still insensitive to their needs for assistance abroad.

- ***The Fear of Traveling***
  Some other interviewees evinced a much dormant fear of traveling, above all after the government restrictions. In sharp contrast with other participants, the consulted persons do not have a bad image of the Government. Contrariwise they take part in a minority that exhibited fear of traveling in the past. This seems to be the case of Jose (male 35 years old) who says:

  "I do not believe that Fernandez should be blamed for this. The problem is that I have a manifest panic to travel after the pandemic. It looks like I will never travel again in the future. What I feel now, is the same I felt after 9/11 (the attacks on the World Trade center). I was almost four years without flying... now I feel the same sentiment of panic in my veins and body."

  Similarly, Rose (female, 45 years old) casts some doubts on the possibility to start a leisure journey in the future:

  "I planned this trip to the US over months I know it is not the best time. We are facing a pandemic! I suffered agoraphobia in the past for what

my therapist suggested to me to make this travel. I do not know when my flight will be placed. I am desperate without money and lodging, I am sleeping at the airport."

This suggests that events like 9/11 or the COVID crisis activate some traumatic experiences which paralyze travelers. This sentiment of aversion seems to remain stable over years until other similar traumatic event takes place.

The effects of COVID-19 in the tourism industry seem to be long lasting. The specialized literature focuses on the opportunity to the rebirth of the tourism industry, as well as the importance of laying the foundations towards a new epistemology, but less is said about stranded tourists, which means the set of tourists who are unable to return to their homes because of COVID restrictions. We have conducted exploratory research finding three indicators: the sentiment of abandonment citizens feel when they are impeded to return because of airspace (borders) closure, a deep process of marginalization that divides the "good" citizens who passively accept the lockdown from those who planned to travel abroad (bad citizens). In this token, participants said to us that they feel discriminated against by the Argentinean government because they traveled to the US. Finally, we come across those who had a type of phobia or aversion to travel. Like in 9/11, the crisis accelerated some dormant fear of flying, a point which merits to be continued in other approaches.

10.6 CONCLUSION

It is difficult to resist the impression that leisure has been captivating scholars and sociologists over the recent decades. From its outset, sociology overlooked leisure as a trivial (naïve) practice focusing the attention on what Emile Durkheim dubbed "the anomie." Although sociology was not interested in leisure as a main object of study, over the years, authoritative voices as Frederic Munne, Norbert Elias, Christ Rojek, and Erich Weber devoted considerable time and efforts to shedding light on the importance of leisure to keep society united. From ancient Romans to the colonial British Empire, leisure practices not only revitalized the social discontent and frustration but also endorsed authority to the imperial law. Having said this, leisure speaks us much of our repressed emotions and expectations without mentioning the global inequalities between human groups. The recent global pandemic of COVID-19 not only grinds the tourism industry and leisure practices to a halt but evinced the material asymmetries and problems of modern global capitalism. With the focus on Argentina, the present manuscript explores the ebbs and flows of what experts known as "vaccine tourism" which means the possibilities or impossibilities of lay citizens to plan holidays to be ultimately inoculated. Vaccine tourism mainly marks a dichotomy between a mobile global elite and the rest of society which remains immobile in Argentina, so to speak, in a strict lockdown. In this vein, leisure inequalities coincide with the previous asymmetries of mobilities (citing Bauman discussing the metaphor between tourists and vagabonds) (Bauman, 1996, 2017).

KEYWORDS

- **COVID-19**
- **leisure inequalities**
- ***marginalized travelers***
- **material asymmetries**
- **Miami**
- ***Otium***
- **SARS-COv2 virus**
- **tourism industry**
- **vaccine tourism**

REFERENCES

Ambagtsheer, F., Zaitch, D., & Weimar, W., (2013). The battle for human organs: Organ trafficking and transplant tourism in a global context. *Global Crime, 14*(1), 1–26.

Balme, M., (1984). Attitudes to work and leisure in ancient Greece. *Greece & Rome, 31*(2), 140–152.

Barbosa, R. B., Costa, J. H., Handayani, B., & Korstanje, M., (2021). The effects of COVID19 in the tourist society: An anthropological insight of the trivialization of death and life. *International Journal of Tourism Anthropology, 8*(2), 179–192.

Baum, T., & Hai, N. T. T., (2020). Hospitality, tourism, human rights and the impact of COVID-19. *International Journal of Contemporary Hospitality Management, 32*(7), 2397–2407.

Bauman, Z., (1996). *Tourists and Vagabonds: Heroes and Victims of Postmodernity*. Vienna, y. (Political Science Series/Institute for Advanced Studies, Dept. Political Science, 30). Vienna: Institute for Advanced Studies (IHS).

Bauman, Z., (2017). Tourists and vagabonds: or living in postmodern times. In: *Identity and Social Change* (pp. 13–26). Abingdon, Routledge.

Béland, D., & Zarzeczny, A., (2018). Medical tourism and national health care systems: An institutionalist research agenda. *Globalization and Health, 14*(1), 1–7.

Bianchi, R., (2022). Tourism, COVID19 and crisis: The case for the radical turn. In: Higgins-Desbiolles, F., Doering, A., & Bigby, C., (eds.), *Socializing Tourism: Rethinking Tourism and Ecological Justice* (pp. 93–108). Abingdon Routledge.

Budiani-Saberi, D. A., & Delmonico, F. L., (2008). Organ trafficking and transplant tourism: A commentary on the global realities. *American Journal of Transplantation, 8*(5), 925–929.

Carr, A., (2020). COVID-19, indigenous peoples and tourism: A view from New Zealand. *Tourism Geographies, 22*(3), 491–502.

Cohen, E., (1972). Toward a sociology of international tourism. *Social Research*, 164–182.

Cohen, E., (1979). Rethinking the sociology of tourism. *Annals of Tourism Research, 6*(1), 18–35.

Cohen, E., (1988). Authenticity and commoditization in tourism. *Annals of Tourism Research, 15*(3), 371–386.

Columb, S., (2019). Organ trafficking: Transplant tourism and trafficking in persons for the removal of organs. *The Sage Handbook of Human Trafficking and Modern-Day Slavery*, 155–173.

Craik, J., (2002). The culture of tourism. In: Rojek, C., & Urry, J., (eds.), *Touring Cultures* (pp. 123–146). Routledge, Abingdon.

Dunning, E., & Rojek, C., (1992). *Sport and Leisure in the Civilizing Process*. Basingstoke, Palgrave Macmillan.

Elias, N., & Dunning, E., (1986). *Quest for Excitement: Sport and Leisure in the Civilizing Process*. New York, Basil Blackwell.

Elkington, S., (2015). Disturbance and complexity in urban places: The everyday aesthetics of leisure. In: Gammon, S., & Elkington, S., (eds.), *Landscapes of Leisure* (pp. 24–40). London, Palgrave Macmillan.

Farzanegan, M. R., Gholipour, H. F., Feizi, M., Nunkoo, R., & Andargoli, A. E., (2021). International tourism and outbreak of coronavirus (COVID-19): A cross-country analysis. *Journal of Travel Research, 60*(3), 687–692.

Garcia-Altes, A., (2005). The development of health tourism services. *Annals of Tourism Research, 32*(1), 262–266.

Goodrich, J. N., (1993). Socialist Cuba: A study of health tourism. *Journal of Travel Research, 32*(1), 36–41.

Gössling, S., Scott, D., & Hall, C. M., (2020). Pandemics, tourism and global change: A rapid assessment of COVID-19. *Journal of Sustainable Tourism, 29*(1), 1–20.

Hall, C. M., (1992). Adventure, sport and health tourism. *Adventure, Sport and Health Tourism*, 141–158.

Hall, C. M., (2012). Medical and health tourism: The development and implications of medical mobility. In: *Medical Tourism* (pp. 19–44). Abingdon, Routledge.

Higgins-Desbiolles, F., Doering, A., & Bigby, C., (2022). Introduction: Socializing tourism: Reimagining tourism's purpose. In: Higgins-Desbiolles, F., Doering, A., & Bigby, C., (eds.), *Socializing Tourism: Rethinking Tourism and Ecological Justice* (pp. 1–20). Abingdon Routledge.

Hunter-Jones, P., (2005). Cancer and tourism. *Annals of Tourism Research, 32*(1), 70–92.

Korstanje, M. E., (2009). Reconsidering the roots of event management: Leisure in ancient Rome. *Event Management, 13*(3), 197–203.

Korstanje, M. E., (2016). *The Rise of Thana Capitalism and Tourism*. Abingdon, Routledge.

Korstanje, M. E., (2017). *Terrorism, Tourism and the End of Hospitality in the West.* Basingstoke, Palgrave Macmillan.

Korstanje, M. E., (2018). The concept of civility and law. In: *Terrorism, Tourism and the End of Hospitality in the 'West'* (pp. 31–55). Palgrave Macmillan, Cham.

Korstanje, M. E., (2018). *The Mobilities Paradox: A Critical Analysis*. Cheltenham, Edward Elgar Publishing.

Korstanje, M. E., (2020). Passage from the tourist gaze to the wicked gaze: A case study on COVID-19 with special reference to Argentina. In: George, B., & Mahar, Q., (eds.), *International Case Studies in the Management of Disasters* (pp. 197–212). Bingley, Emerald Publishing Limited.

Korstanje, M. E., (2021). Tourism disaster preparedness post-COVID-19 pandemic: The example of Argentina. In: Gowreesunkar, V., Maingi, S. W., Roy, H., & Micera, R., (eds.), *Tourism Destination Management in a Post-Pandemic Context* (pp. 69–84). Bingley, Emerald Publishing Limited.

Korstanje, M., & George, B., (2021). *Mobility and Globalization in the Aftermath of COVID-19: Emerging New Geographies in a Locked World.* Basingstoke, Palgrave Macmillan.

Korstanje, M., & George, B., (2022). *The Nature and Future of Tourism: A Post COVID19 Context.* Palm Bay, Apple Academic Press.

Krippendorf, J., (2010). *Holiday Makers*. Abingdon, Routledge.

Kyle, D. G., (2012). *Spectacles of Death in Ancient Rome*. Abingdon, Routledge.

Lash, S. M., & Urry, J., (1993). *Economies of Signs and Space* (Vol. 26). London, Sage.

MacCannell, D., (1973). Staged authenticity: Arrangements of social space in tourist settings. *American Journal of Sociology, 79*(3), 589–603.

MacCannell, D., (1976). *The Tourist: A New Theory of the Leisure Class*. Berkeley, University of California Press.

MacCannell, D., (2002). *Empty Meeting Grounds: The Tourist Papers*. London, Routledge.

Montgomery, H., (2010). *Child Sex Tourism*. Oxford, Goodfellows.

Mostafanezhad, M., Cheer, J. M., & Sin, H. L., (2020). Geopolitical anxieties of tourism: (Im) mobilities of the COVID-19 pandemic. *Dialogues in Human Geography, 10*(2), 182–186.

Munné, F., & Codina, N., (2002). Ocio y tiempo libre: Consideraciones desde una perspectiva psicosocial. (Leisure and free time: Some questions from a social psychology perspective). *Revista Licere-Brasil, 5*(1), 59–72.

Munné, F., (1980). *Psicosociología del Tiempo Libre: Un Enfoque Crítico (Psychosociology of Leisure: A Critical Approach)*. Trillas, Mexico.

Prideaux, B., Thompson, M., & Pabel, A., (2020). Lessons from COVID-19 can prepare global tourism for the economic transformation needed to combat climate change. *Tourism Geographies, 22*(3), 667–678.

Reisman, D. A., (2010). *Health Tourism: Social Welfare Through International Trade*. Cheltenham, Edward Elgar Publishing.

Riesman, D., Glazer, N., & Denney, R., (2001). *The Lonely Crowd*. New Haven, Yale University Press.

Ritzer, G., (2007). Inhospitable hospitality? In: *Hospitality: A Social Lens* (pp. 143–154). Abingdon, Routledge.

Rogerson, C. M., & Rogerson, J. M., (2020). COVID-19 tourism impacts in South Africa: Government and industry responses. *GeoJournal of Tourism and Geosites, 31*(3), 1083–1091.

Rojek, C., (2005). An outline of the action approach to leisure studies. *Leisure Studies, 24*(1), 13–25.

Rojek, C., (2005). Leisure theory. *Principles and Practice*. New York, Springer.

Rojek, C., (2013). *Capitalism and Leisure Theory (Routledge Revivals)*. Abingdon, Routledge.

Romagosa, F., (2020). The COVID-19 crisis: Opportunities for sustainable and proximity tourism. *Tourism Geographies, 22*(3), 690–694.

Rose-Redwood, R., Kitchin, R., Apostolopoulou, E., Rickards, L., Blackman, T., Crampton, J., & Buckley, M., (2020). Geographies of the COVID-19 pandemic. *Dialogues in Human Geography, 10*(2), 97–106.

Rowe, D., (2002). Producing the crisis: The state of leisure studies. *Annals of Leisure Research, 5*(1), 1–13.

Ryan, C., & Kinder, R., (1996). Sex, tourism and sex tourism: Fulfilling similar needs?. *Tourism Management, 17*(7), 507–518.

Salazar, N. B., & Graburn, N. H., (2014). *Tourism Imaginaries: Anthropological Approaches*. Oxford, Berghahn books.

Sigala, M., (2020). Tourism and COVID-19: Impacts and implications for advancing and resetting industry and research. *Journal of Business Research, 117*, 312–321.

Smith, C., & Jenner, P., (2000). Health tourism in Europe. *Travel & Tourism Analyst*, (1), 41–59.

Spracklen, K., (2011). *Constructing Leisure: Historical and Philosophical Debate*. Basingstoke, Palgrave Macmillan.

Spracklen, K., (2013). *Whiteness and Leisure*. New York, Springer.

Stebbins, R. A., (2011). Leisure studies: The road ahead. *World Leisure Journal, 53*(1), 3–10.

Suess, C., Baloglu, S., & Busser, J. A., (2018). Perceived impacts of medical tourism development on community wellbeing. *Tourism Management, 69*, 232–245.

Theobald, W. F., (2005). The meaning, scope, and measurement of travel and tourism. *Global Tourism, 3*, 23–48.

Tribe, J., (2010). Tribes, territories and networks in the tourism academy. *Annals of Tourism Research, 37*(1), 7–33.

Trunfio, M., (1991). A theological perspective on the ethics of leisure. *Leisure and Ethics: Reflections on the Philosophy of Leisure*, 151–59.

Tzanelli, R., & Yar, M., (2020). Atmospheres of the inhospitable in staged kidnappings. *Consumption Markets & Culture*, 1–17.
Tzanelli, R., (2016). *Thanatourism and Cinematic Representations of Risk: Screening the End of Tourism*. Abingdon, Routledge.
Urry, J., (1990). The consumption of tourism. *Sociology, 24*(1), 23–35.
Urry, J., (2002). *The Tourist Gaze*. London, Sage.
Veblen, T., (2017). *The Theory of the Leisure Class*. Routledge, Abingdon.
Weber, E., (1969). *El Problema del Tiempo Libre. (The Problems of Leisure)*. Editorial Nacional, Madrid.
Wilson, J., (1980). Sociology of leisure. *Annual Review of Sociology, 6*(1), 21–40.
Yar, M., & Tzanelli, R., (2019). Kidnapping for fun and profit? Voluntary abduction, extreme consumption and self-making in a risk society. *Hospitality & Society, 9*(2), 105–124.

CHAPTER 11

Ethics, Post-Disaster Marketing, and COVID-19

MAXIMILIANO E. KORSTANJE[1] and HUGUES SÉRAPHIN[2]

[1]*Department of Economics, University of Palermo, Larrea ST 1079, Buenos Aires, Argentina*

[2]*Oxford Brookes Business School, United Kingdom*

ABSTRACT

This chapter has focused on the problem of disaster marketing and management as well as its impact on the currently existing literature. Disaster marketing seems to operate under the premise that disasters are impossible to prevent. In this vein, policymakers should adopt resilient programs that mitigate the impacts of disasters once they take a hit. This academic position recently emerged just after 2010 and widely gained traction in the academy. The chapter discusses practical applications of disaster marketing to a post-COVID-19 world.

11.1 INTRODUCTION

There is an interesting point of entry is given in the constellations of disaster studies respecting the importance of technology in mitigating or creating global risks. As Ulrich Beck (1992) noted, technology may help human development, but paradoxically it may lead towards an apocalyptic future. He centers the analysis after Chernobyl nuclear accident, in which case, as he adds, the same technology, which was oriented to

Post-Disaster and Post-Conflict Tourism (An Updated 2nd Edition), Maximiliano E. Korstanje, Hugues Seraphin, and Vanessa G. B. Gowreesunkar (Eds.)

protect humanity, paves the way for the rise of a nuclear apocalypse. From Beck's days onwards, in the academy, some leading voices have alerted on the negative effects of climate change which accelerated an escalation of disasters that may harm underdeveloped economies (Giddens, 2009, 2011; Urry, 2015; Doyle, 2007). Quite aside from this, experts agree that Western narratives enthusiastically emphasize the importance of adaptation to mitigate the impacts of man-made disasters. In so doing, the structural background that facilitated the conditions of the disaster are untouched. To some extent, the notion of adaptation does not resolve the factors behind the event (Chaturvedi & Doyle, 2016; Doyle & Chaturvedi, 2010). In this context, post-disaster marketing helped not only officials and the net of experts but contributed victims to commoditize the disaster as a form of product, to be consumed by others. Over years, scholars toyed with the belief that tourism would be affected by disasters, but recently they are changing to think tourism commoditizes disasters given new forms of consumptions (Ladking et al., 2008; Walters & Mair, 2012; Korstanje & Ivanov, 2012; Séraphin, Butcher, & Korstanje, 2016; Séraphin & Gowreesunkar, 2017).

The present chapter deals with a difficult problem for marketing designers and policymakers that today marked the agenda of many governments and nations in a post-COVID-19 context. Whilst tourism benefited considerably many domestic economies, no less true is that the degree of dependency the industry generates may be a serious problem for underdeveloped economies. The turn of the century brought some global risks such as climate change, terrorism, natural disasters and so forth, which places the industry of tourism and hospitality in jeopardy. In this context, marketing-related specialists devoted considerable attention to understanding furtherly how the mechanism of resiliency operates. Veering from an individual towards a social dimension, resilience played a vital role in revitalizing the psychological frustrations of survivors in post-disaster contexts. Equally important, tourism seems to be an interesting conduit to overcome traumatic experiences and accelerating the times in the process of recovery. In this vein, though the term resiliency was originally coined by Victor Frankl after WWII (for further details see Wong, 2014), less attention was given in the fields of tourism and hospitality. This chapter not only aims to fill this gap, placing the specialized literature under the critical lens of scrutiny, but also poses a new discussion respecting '*post-disaster marketing*,' which merits to be continued in the years to come.

Whilst Post-disaster marketing offers a fertile ground to implement plans of contingency, which not only saves lives, but it is also important not to lose sight of the reasons behind disasters. At the same time, the lessons of disasters are commoditized to be grasped by global spectatorship, serious risks the disaster repeats turn higher. Our viewpoint frames within the fields of "ethical marketing."

11.2 TOURISM, SOCIETY, AND RESILIENCE

Starting from the premise that disasters are socially determined and culturally negotiated, Enrico Quarantelli (2005) says that what is seen as a tolerable risk in one context is not in another. Disaster studies are fraught with countless fragmented knowledge which is based on different definitions and interpretations about what a disaster means. As Quarantelli puts it, disasters cannot be determined or estimated by the approximate material losses or casualties. What is perceived as threatening for a community is normal in another. For Quarantelli, disaster studies lay the foundations for a new discipline that gives the conceptual framework to define and grasp disasters. The communication process plays a leading role in disaster studies. Over the recent decades, the mass media occupied a central position not only packaging information about disasters but also creating a spectacle around them, oriented to entertain the masses (Korstanje, 2010; Chaturvedi & Doyle, 2016).

In her book, "*The Shock Doctrine: The Rise of Disaster Capitalism,*" Naomi Klein (2007) explains that neoliberalism, far from correcting the socio-economic background that facilitated disasters, imposes to citizens policies otherwise would be rejected. The system, as Klein notes, never corrects the glitches behind the disaster rather it potentiates its effects while recycling the affected zones. Disasters are the main commodities of capitalism. In consonance with this idea, Jean Baudrillard argues convincingly that in global capitalism the future not only interrogates the agency but also creates what he dubbed as pseudo-events. Based on the Plot of Minority Report, which means a dystopian world where criminals are caught by the police earlier the crime is committed, he says overtly that society inflates some global risks which never happen in reality, as least as they were originally imagined (Campbell, 2010; Korstanje, 2013). In this respect, Haigh & Amaratunga (2010) called attention to the need of integrating the disaster mitigation literature with resiliency. Whilst disasters often obstruct the correct functioning of society, resiliency optimizes not only the timeframe but also the recovery process notably. As authors write:

> *"In recognition of the devastating and long-term consequences that can result from a disaster, the term resilience has been widely adopted by researchers and policy makers in an attempt to describe how they like to reduce our society's susceptibility to the threats posed by natural, human, and technical hazards (Haigh & Amaratunga, 2010, p. 13)."*

As this argument is given, disasters cannot be evaluated or judged by their immediate effects, but through the manners, they are internalized by survivors (Keane, 1996). Disasters often take a hit altering the emergency-led protocols while igniting a sentiment of panic in society, a sentiment which sometimes persists in time. The post-disaster recovery facet is accompanied by ideological discourses oriented to blame some ethnic minorities for material losses or casualties (Lacy, 2001). These types of discourses are embodied in specific practices crystallized in dark tourism sites. Dark tourism can be seen as an emerging sub-discipline that studies the allegories and discourses revolving around spaces of mass destruction or death. These practices, which derive from traumatic experiences, are culturally enrooted in the territory (Johnston, 2013). Some voices allude to dark tourism practices as forms of heritage codified to re-interpret a space of trauma. In so doing, dark tourism sites are human attempts to domesticate death (Poria, 2007; Raine, 2013). As Erik Cohen eloquently alerts, traumatic death divides the world in two, the victims and survivors. He introduces the term populo-site to define the contours of territory where the memory is fabricated. Whilst there are a lot of sites that re-memorize the traumatic event, only a few – if not only one – provides an authentic experience to visitors. For Cohen, populo-sites are the place where everything started, where the first bodies fell, or the founding hero is buried. He ascertains that dark tourism serves as a pedagogical form of learning for a similarly minded event to be avoided in a no so-distant future. This conceptual position is continued by White & Frew (2013) who hold the thesis that dark consuming sites are politically manipulated to share a story or a message which is negotiated with the community. Far from being unilateral, these stories are internalized, rejected or negotiated by community members. Dark sites offer fragmented interpretations of the traumatic event, but under some conditions, the message consents while going in a one-side direction. In her seminal book "*Heritage That Hurts*," Sather Joy Wagstaff (2016) confronts the emotionality, as well as the sentiment of reciprocity, manifested just after a disrupting event. Death makes not only humans equal but also actives the necessary resilient mechanisms supporting the community as long as the post-disaster recovery facet. The sentiment of solidarity awakened up by death is ultimately manipulated by the status quo to keep its privileged position. As she explains, the nature of heritage corresponds with a political distortion of the reasons behind the event. The emotional logic of dark tourism consumption is certainly given by conceptual distortions organized to legitimate the ruling elite. Phillip Stone (2011, 2012) understands dark tourism as an anthropological attempt to theorize about the "Other's death," to create a synthesis between visitors and the alterity. Dark tourists are psychologically moved to understand their finitude through the "Other's death." Some critical studies stress the ambiguous logic of dark tourism consumption, which often makes the "Other's pain" a main commercial commodity to be exchanged. For these voices, visitors – far from being sensitive—are in quest of authentic experiences that lead them to maximize their pleasure while affirming their status (Bowman & Pezzulo, 2009; Hartmann, 2014; Tzanelli, 2016; Korstanje, 2016; Carrigan, 2014). This radical "turn" is mainly marked by a critical position on dark tourism studies as well as the methodological limitations of applied research. Dark tourism studies are based on formal or informal questionnaires or interviews conducted with visitors to dark sites. This methodological stance ignores that sometimes interviewees lie to protect their interests or simply are unfamiliar with their emotionality (Korstanje, 2016). Maria D. Buda tries to conciliate dark tourism heritage theory with a radical position. She combines the advances of psychoanalysis – above all the Freudian concept of death drive – with cultural studies. With a focus on the theory of voyeurism, Buda & McIntosh hold the thesis that dark tourism exhibits a much deeper desire aimed at disciplining the dangers which are perceived and projected by the inner world of visitors (Buda & McIntosh, 2013). This means that everyone finds in these types of mourning places, exactly what it is looking for. Still further, the death drive in tourism studies represents an interesting point of investigation that explains the predilection of visitors for consuming post-disaster environment (Buda, 2015a, b).

Last but not least, Shondell Miller (2008) and Shondell Miller, Gonzalez, & Hutter (2017) coined the term '*disaster-tourism*' to describe the demands for visiting spaces of mass destruction, death, and mourning in post-disaster contexts. As a vehicle towards reconstruction, disaster tourism allows rapid recovery and further financially-aid programs for affected communities. Hence post-disaster marketing offers a fertile ground to the explorations of new horizons of research, which made pain its main commodity.

11.3 POST-DISASTER MARKETING IN TOURISM

As a newly emerging discipline, post-disaster marketing recently captivated the attention of some scholars in the fields of tourism and hospitality (Gotham, 2008; Chew & Jahari, 2014; Hajibaba et al., 2015; Mair, Ritche, & Walters, 2016; Tzanelli & Korstanje, 2016). The specialized literature presents marketing and more specifically the branding process as a vital force towards more sustainable and competitive destinations (Blain, Levy, & Ritchie, 2014; Park, Eisingerich, Pol, & Park, 2013; Park & Petrick, 2006). At a closer look, 'country image and destination image influence revisit intention' (Zhang, Wu, & Buhalis, 2017, p. 1). A positive brand image also positively contributes to increase the export of a destination and encourage investments (Vitic & Ringer, 2008). It also contributes 'to attract higher-spending tourists (Park & Petrick, 2006). In other terms, the negative organic destination image seriously affects the tourism industry as well as the visitors' numbers. Taking the example of Haiti (a postcolonial, post-conflict, and post-disaster destination with a negative image), Hugues Séraphin (2014) explores how the negative destination image (characterized by high levels of political instability and poverty) impedes Haiti to situate as a leading destination in the Caribbean. The 2010 earthquake further flattened the image and performance of the destination. A post-disaster marketing strategy and more importantly, a repositioning of the destination appeared as a 'must' as it allows the destination to communicate on the positive image the destination already held; correct negative images or create a new image (Chacko & Marcell, 2008). Post-disaster or post-conflict marketing is also an opportunity for a destination to reinvent itself. That was for instance the case of Montenegro after the disintegration of Yugoslavia (Vitic & Ringer, 2008). A panel of options is available to destinations to rebrand themselves after a disaster or a conflict. Among other things, the Haitian destination marketing organization (DMO) rebranded the destination by changing the DMO logo (Séraphin, Ambaye, Gowreesunkar, & Bonnardel, 2016). Because of the role of the key logo in destination branding (Kladou, Kavaratzis, Rigopoulou, & Salonika, 2016), the change of logo is quite a common strategy adopted by destinations after a disaster (whether natural or not). As for the New Orleans' DMO and DMCs (and other stakeholders), after hurricane Katrina, they use a variety of methods. Among these are: Media coverage; visitors who had already visited the destination were encouraged to revisit; endorsement from celebrities (who urged people to visit the destination); DVDs showing the assets of the destination; online strategy; and variety and quality of facilities of the destinations were put forward; etc. (Chacko & Marcell, 2008). As for Séraphin, Butcher, & Korstnaje (2016), they suggested a less traditional post-disaster/post-conflict marketing strategy, namely the education of tourists at the pre-visit stage using Visual Online Learning Materials (VOLMs).

Current issues in Europe and the world are likely to see more post-disaster and post-conflict marketing strategies appearing. These strategies are also likely to be more innovative due to the nature of the issues. Let's consider Europe first. Across many European destinations (Barcelona, Cambridge, Dubrovnik, Florence, Oxford, Rome, Stratford-on-Avon, Venice, and York) anti-tourist protesters are voicing (protests, graffiti, and physical intimidation) their concern regarding the development of the tourism industry. These destinations under anti-tourist anger are already planning to monitor tourists and tourism more closely either by limiting the number of visitors; limiting cruise ships; introducing new tourism taxes and fines; having special patrols; encouraging tourists to visit other parts of the country that are less-visited; restricting some places to tourists; ensuring that tourism is an enriching experience for visitors and hosts alike; encouraging tourists to visit beyond the central sights; diversifying tourist activities; reducing seasonality and addressing the needs of the local community (Coldwell, 2017; Tapper, 2017). The reasons for this rise in anti-tourism are because the large number of tourists visiting some destinations is putting the UNESCO World Heritage status at risk; tourists are affecting the quality of life of locals and particularly killing neighborhoods; putting at risk the sustainability of the tourism industry of the destination (Leadbeater, 2017); the tourists are not respecting the cities, history, arts, and inhabitants by defacing the surroundings, dumping thrash, buying counterfeit goods, sitting anywhere and spending very little money particularly day-trippers (Buckley, 2017; Leadbeater, 2017). For Tapper (2017), a tipping point has been reached. On that basis, we can assume that *trexit* (short for tourism exit) might become an option that some destinations or at least some anti-tourist movement would consider. *Trexit* is not a farfetched scenario. Indeed, despite the importance of the tourism industry for many destinations, it seems that the locals are now ready to sacrifice the benefits of the industry for a better quality of life. In this chapter, we believe that this new trend toward anti-tourism is to be associated to bigger trend associated with the terms: 'Quality,' 'healthy' and 'natural.' In agri-food, this could be exemplified by the popularity of organic food. Among other things, this new trend means that farmers are not using chemical with the consequence of a lesser production, which means less income for them. Tourism seems to go toward that way (quality instead of quantity). A *trexit* option would be damageable for the image and economy of the destinations which will be going for this option. What post-conflict marketing strategy are they going to adopt to revamp their image?

Now let's consider the Caribbean. On Tuesday 02.09.17, hurricane Irma severely damaged the Caribbean. The negative impacts of this hurricane were particularly visible in Saint-Barthelemy, Saint-Martin, Barbuda, and Anguilla. For instance, in Saint-Martin, 95% of the locals lost their house and 50 people or so were injured. Destinations also had to plan the evacuation of tourists. In Cuba, it is some 10,000 tourists were evacuated (France-Antilles, 2017), etc. What post-disaster marketing strategy is the Caribbean going to adopt to revamp its image? This is all the more important as Holder (1980) explained that tourism can play a major role in the economic development of the Caribbean and should therefore be a priority. Also, for some islands, the industry is the main, even only source of income (Peter, 1980).

Post-disaster or post-conflict marketing strategy for a destination is not an option but a sine-qua-non condition for survival. That said, there are some limitations to this approach so far presented as extremely important for destinations after a disaster (or a conflict). The first limitation arises from the fact that DMOs strategies after conflicts and/or disasters are more geared toward economic recovery than the well-being of the local population. Taking the example of Haiti, Séraphin, Gowreesunkar, &

Ambaye (2016) explain that the destination should have considered the improvement of the condition of the life of the locals (well-being; education) alongside the rebranding of the destination and all the effort made to attract international firms. A change of paradigm needs to be operated. DMOs need to be more inward centered (local population). Local people are important for the sustainability of the tourism industry of a destination. The second limitation of post-disaster or post-conflict marketing strategy is based on the Janus-faced character of the tourism industry (Sanchez & Adams, 2007). Still taking the example of Haiti, Séraphin et al. (2016) explains that in the process of rebranding the destination, the current logo of the destination overlooked the importance of symbols telling the narrative of the destination, meaning that the current logo lost its authenticity. When we know the importance of authenticity in terms of the ability of a destination to attract tourists, this marketing strategy defeated its objectives. Last but not least, most post-disaster or post-conflict marketing strategies seem to overlook the fact that tourists who visited a destination before a conflict or disaster, would keep in mind a positive image of the destination if their experience there (before the conflict or disaster) was good (Hubner & Gosling, 2012). Marketing strategies are sometimes reactionary instead of being thought through.

Post-disaster marketing also meets with some ethical and conceptual limitations which deserve to be first discussed. Hammond & Wellington (2013, pp. 59, 60) define ethics as a moral principle or a code of conduct that governs what people do. It is concerned with the way people act or behave.' Lovelock & Lovelock (2013) added the fact that there is a call for the tourism industry to be more ethical in the way it does business. More specifically, the industry should be beneficial to all stakeholders and contribute to environmental, social, and economic goals. First and foremost, there is an epistemological flaw in these types of studies, which over-valorizes the view of tourists as the precondition of what the marketing model would propose. Though this works in normal conditions, disasters open the doors to outstanding situations, where victims are ethically subordinated to the tourist's whim. Some critical studies conducted under the auspices of gentrification theory reveals that after disasters the material asymmetries that characterized the lifestyle in society are enlarged. To wit, the international investors expropriate the best lands pushing the affected citizens to the periphery. As a result of this, thousands of neighbors lost their original homes and are limited to soils with limited resources (Carter et al., 2007; Aldrich, 2012; Korstanje, 2014). To correct this, the government should intervene in the design of the post-recovery process more actively whilst sometimes foreign capitals do not recognize the domestic laws. As this backdrop is given, a second problem lies in the fact that dark sites suggest that there is a previous political intervention in packaging and disseminating the message, which often is aimed at tergiversating the real reasons of disaster. When this happens, the community not only is unfamiliar with the background but also will be prone to continue with the same behavior that led to the catastrophe. Last but not least, whereas – as a leading discipline – marketing devotes resources to mitigate the effects of disasters it perpetuates the pain of victims as the main commodity to exchange. Bianca Freire Medeiros, a Brazilian colleague, studied slumming in Brazil embracing enthusiastically the thesis that slum-tourism is paradoxically oriented to replicate the conditions of exploitation that made from poverty a criterion of attraction for other external visitors. This represents a serious problem simply because slum tourism, far from reducing poverty replicates its economic and symbolic background (Freire Medeiros, 2014). The same may very well be objected to post-disaster marketing.

The Janus-faced character of the tourism industry and the limitation of post-disaster and/or post-conflict marketing strategy are calling for an ambidextrous management approach. 'Ambidextrous' refers to the addressing of two antithetical goals simultaneously (Smith, 2017; Mihalache & Mihalache, 2016): innovating and promoting the destination on the one hand *as it is*, yet, on the other hand, innovating towards greater ethics *as it could be*. This approach is presented by Smith (2017) as important for the survival of any organization in both the short-term and long term. Last but not least, post-disaster marketing seems not to be familiar with untypical events like COVID-19 virus outbreak. Although some studies caution on the risks of similar virus outbreak (like Ebola or SARS) posited on the tourism industry in the past, nothing prepared experts for COVID-19. The industry is the main carrier and victim of these types of outbreaks. The high mobilities culture, associated to the expansion of transport means and the urban over-crowding, disseminates a lethal virus in a question of days (Wilder-Smith, 2006; Gu & Wall, 2006). It is difficult not to resist the impression that COVID-19 generated long-lasting and devastating effects on the tourism industry and travel behavior. The strictest measures and dispositions emanated from authorities to control the tourist fluxes failed. COVID-19 infected more than 230 million whilst killed more than 5 million persons. As a real catastrophe, COVID-19 introduces radical transformation in travel behavior which includes new regulatory requirements, travel bans, health passports and the temporal closures of borders and airspace without mentioning the hostility against Chinese tourists and the geopolitical tensions (Korstanje & George, 2021, 2022). Beyond this apocalyptic future, innovative forms of dark tourism practices surface. In the next section, we discuss the implications of pandemic tourism, which means a niche (formed by consumers) psychologically motivated to visit countries, or cities widely characterized by high viral circulation.

11.4 A NEW REFRESH AFTER COVID-19 PANDEMIC

Over the recent years, in the mid of COVID-19 pandemic, many young tourists opted to travel to Wuhan, China, where the first cases were originally reported. This new phenomenon opened a hot debate revolving around the genesis of a new niche, pandemic tourism. Unlike vaccine tourism, a term developed in earlier chapters, pandemic tourism refers to the motivation to travel to cities or places in quarantine or seriously affected by a lethal disease. The word pandemic denotes a presence of an infectious disease most probably resulted from a virus outbreak which cannot be controlled or contained by health authorities. Pandemics are expanded, encompassing several nations and the four continents (Yang, Zhang, & Chen, 2020). The term pandemic tourism is used to denote a decision or curiosity to travel to risky destination seriously harmed by COVID-19 and other virus outbreaks. Doubtless, pandemic tourism can be framed as a sub-type of dark tourism consumption. Visitors look to get authentic and outstanding experiences to be shared with peers and the global audience. We have made a short inquiry sampled on 35 participants (from 25 to 55 years old) separated among 15 males and 20 females. Participants are Argentinean tourists who traveled abroad during the pandemic and travel bans. The sample was placed in two groups: those who have been stranded abroad by tourism or for working purposes (i.e., cruise-workers, postgraduate students, businessmen, or pilots), and those who voluntarily started a journey defying the cautions issued by health authorities. Both were named as cluster A and cluster B. Because of confidentiality issues, real names of interviewed people have been changed.

Because of the lockdown still remains, personal meetings – or face-to-face interviews—are not allowed in Argentina. Hence, the used method was snowball while the interviews were conducted through Zoom and Skype.

As the previous backdrop, both samples have serious differences and commonalities. Participants wholly manifested a much deeper fear of being infected by COVID during their travels. This suggests not only the impossibility of receiving a correct medical treatment but also the coherent follow-up once returned. Claudio (male, 27 years old) is flight attendant and says:

> *"I have a great experience in traveling and particularly I am not afraid to be sick out of Argentina, but in this case, I was terrified to think in the odds of being infected by COVID-19. If you ask me, hospitals and health system in the UK suddenly collapsed and the information was poor, we know nothing when and under what conditions we would return to home."*

A similar story is found in Maria (female 39 years old) who works as a historian and was surprised while attending to a Conference in Spain:

> *"I was surely terrified when COVID-19 spread; we had neither information to the details of flight to Buenos Aires nor any idea if the government will open the airspace to allow our repatriation. What is worse, all this resembles a horror history and apocalyptic genre where the bottom-days are around the corner."*

In sharp contrast to panic sentiment, those who have flown after the lockdown showed a manifest indifference to "Coronavirus decease" which was accompanied by a political hostility against President Fernandez. Carlos (male, 37 years old), who works as an agricultural engineer, traveled to Paris accompanied of his sons:

> *"I have no fear, this is an invention of China and the media to paralyze the industry in the West, this is like little flu which is being magnified by journalism. I do not share the decision of President Fernandez to close the airspaces, I have rights, and have the right to move as I want, from*

here to there, he is a dictator and this virus allowed him to impose a regime in my country. I made the decision to travel anyway, and I am proud of that, this is an act of resistance!"

The worries for losing the job are the second significant element in discussion in the testimonies of participants. Some interviewed participants in cluster A evinces a manifest concern of being laid off or fired once returned to the home. This sentiment is particularly observed in tourism or hospitality workers. As Joaquin eloquently stresses:

"Unemployment is one of my greatest fears, I am a tour operator in Buenos Aires, and COVID-19 evinces the end of tourism. The lockdown not only limits public circulation but also mobilities and tourism. The system should face a serious crisis, probably like we have in 2001 in Argentina. Thousands and Thousands of Argentineans will impoverish because of COVID-19. The government should assist rapidly to the industry and implement efficient policies of recovery. The enemy is not the virus but poverty."

Rather, in cluster B, the worries stem from the possibilities to weaken the democratic institutions taken advantage of the fear, the crisis generates. Fabiana (female, 27 years old) utters:

"I am afraid of the Argentine government, not the virus. The fear, this government is manipulating, facilitates things to undermine the check and balance forces that makes democracy strong, Fernandez wants to impose fear and the lockdown because his policies not to be widely rejected. Like in the case of terrorism, where the "terrorist" is demonized to impose economic policies that in other circumstance will be rejected, now Fernandez is adopting the lockdown to impoverish the private sector and in this way, introducing Chavism expropriating private fortunes, and properties (like in Venezuela now in Argentina)."

Both testimonies agree poverty caused by the economic crisis is an important factor of COVID-19, but they differ in their reactions. Fabiana adopts a partisan (if not ideological) position blaming President Alberto Fernandez by the crisis, whilst Joaquin signals to the risks of collapse in the tourism industry as well as service sectors. For something is in danger. Joaquin alludes to the economic factor, while Fabiana signals to the fragility of the democratic institutions.

The need of getting unique or authentic experiences – as literature remarks—is a third factor to be kept into consideration. Like in post-disaster or dark tourism, there is a segment of thrill-seekers who move to zones of disaster, war, or conflict only to feel outstanding experiences. In cluster A, almost 10 participants answered that they were suddenly trapped after March and never would fly abroad after the lockdown while 12 in Cluster B were moved to start a trip with the end of understanding the real effects of COVID-19 in the world. This seems to be the case of Juan (male, 45 years old), a sociologist, who acknowledged that:

> *"COVID-19 offers an apocalyptic scenario where things will not be normal any longer, this is an outstanding time, I loved to live. I am not frightening to travel abroad, I needed to be closer to this dark experience I needed to feel the adrenaline to be in destinations highly infected by COVID-19. My first destination was China, preferably Wuhan but it was impossible, so I bought a ticket towards New York. I am not an unconscious guy. This historical event should be lived with great intensity!"*

Instead, Adriana (postgraduate student) offered a different viewpoint:

> *"these are sad days, I was studying anthropology in Australia when this happens. I thought I would never come back to my country, to be with my parents and sisters. This event will change me forever I will not see the travels and tourism, in the same way, I loved to travel and study worldwide, understanding other cultures and nations, be in contact with different people. Now, I have suffering insomnia, and some irrational phobias to take a bus or a train. I stay at home now, and probably will be a long time without planning holidays."*

Another interesting sub-cluster, which we have not typified in this chapter is reserved for the expatriates – which mean professionals – who even if working abroad—opted the decision to come back to Argentina. This is the case of Emanuel who is the hotel's concierge in Portugal. He gives a romantic answer when this interviewer asked for what he felt once expatriated.

> *"Difficult to say I am sad, the feelings in my side are mixed... I decided to study tourism in Europe and this goal was successfully achieved, but COVID-19 allowed me to come back to my home. Believe me, though you work hard, you will never be a European guy. An expatriate feels a half side is emptied, you are happy by your achievements but here is not your home, they do not speak your tongue, they do not think like you, after all, you will always be a stranger. I note my compatriots are very tough when they return to Argentina. They come back temporarily by holidays or to visit friends and take a lodge at a hotel or rent a car. They scornfully treat badly to the staff, often comparing the service in a third world country with first world ones. To your question, I must confess, now I reckon they are hostile to them because they hate they will never be native of the society they admire, their half remains emptied. I am complete and it is a merit of COVID-19."*

To sum up, and in an exploratory dimension, the gathered preliminary results evince promising hypothesis which may be very well continued in a future approach. However, these outcomes cannot be extrapolated to other universes and situations. The obtained outcome permits to draw three different grids or psychological characters, which are based on the factors that impact to make the decision to travel in a pandemic context:

1. **Stranded Tourists:** This segment represents all tourists who were involuntarily trapped by the global lockdown administered by nation-state because of COVID-19. These tourists develop a "precautionary character."
2. **Pandemic Seekers-Deniers:** This segment is mainly drawn by those tourists who motivated to defy the government or to be closer to the infected epicenter, make the decision to travel abroad in spite of the imposed travel-bans. These tourists develop a "defying character."
3. **Nostalgic Expatriates:** The segment is formed by expatriates who work abroad but are emotionally or economically forced to return home. These tourists emotionally instill a "relational character."

11.5 CONCLUSION

In the present chapter, we devote our time and efforts to deciphering the motivation of dark or post-disaster tourists. Since dark tourists often look something different visiting sites of mass destruction of death no less true are that they want to bring some reflection on their finitude through the "Other's death." Over the recent years, experts alert on the rise of a new segment of tourists who put their lives at risk to be in zones of conflict, war, or disaster. The chapter discusses the advantages and disadvantages of post-disaster consumption while laying the foundations for a new understanding of ethics. Pandemic tourism seems not to be an exception. Thousands of tourists arrange their travel to fly China or another country putting their lives in jeopardy. To some extent, the real consequences of COVID-19 in the tourism industry remain diffuse even for specialists. Pandemic tourism ignites a debate around the notions of security, curiosity, or risk, and of course, in so doing, it sheds light on the importance of post-disaster tourism in the years to come. Post-disaster tourism, like vaccine tourism, contributes positively not only to the tourism industry recovery but in building resilience as a form of social capital.

KEYWORDS

- **adaptancy**
- **COVID-19**
- **disaster marketing**
- **disasters**
- **new real**

REFERENCES

Aldrich, D. P., (2012). *Building Resilience: Social Capital in Post-Disaster Recovery*. Chicago: University of Chicago Press.

Baudrillard, J., (2013). *The Intelligence of Evil: Or THE Lucidity Pact*. London: Bloomsbury.

Beck, U., (1992). *Risk Society: Towards a New Modernity* (Vol. 17). London: Sage.

Blain, C., Levy, S. E., & Ritchie, J. R. B., (2014). Destination branding: Insights and practices from destination management organization. *Journal of Travel Research, 43*, 328–338.
Bowman, M. S., & Pezzullo, P. C., (2009). What's so 'dark' about 'dark tourism'?: Death, tours, and performance. *Tourist Studies, 9*(3), 187–202.
Buckley, J., (2017). *Florence Launches Campaign Telling Tourists How to Behave*. Available from: http://www.cntraveler.com/story/florence-launches-campagin-telling-tourists-how-to-behave (accessed on 20 January 2022).
Buda, D. M., & McIntosh, A. J., (2013). Dark tourism and voyeurism: Tourist arrested for "spying" in Iran. *International Journal of Culture, Tourism and Hospitality Research, 7*(3), 214–226.
Buda, D. M., (2015). *Affective Tourism: Dark Routes in Conflict*. Abingdon: Routledge.
Buda, D. M., (2015). The death drive in tourism studies. *Annals of Tourism Research, 50*, 39–51.
Campbell, E., (2010). The future (s) of risk: Barthes and Baudrillard go to Hollywood. *Crime, Media, Culture, 6*(1), 7–26.
Carrigan, A., (2014). Dark tourism and postcolonial studies: Critical intersections. *Postcolonial Studies, 17*(3), 236–250.
Carter, M. R., Little, P. D., Mogues, T., & Negatu, W., (2007). Poverty traps and natural disasters in Ethiopia and Honduras. *World Development, 35*(5), 835–856.
Chacko, H. E., & Marcell, M. H., (2008). Repositioning a tourism destination. *Journal of Travel & Tourism Marketing, 23*(2–4), 223–235.
Chaturvedi, S., & Doyle, T., (2016). *Climate Terror: A Critical Geopolitics of Climate Change*. Basingstoke: Springer Nature.
Chew, E. Y. T., & Jahari, S. A., (2014). Destination image as a mediator between perceived risks and revisit intention: A case of post-disaster Japan. *Tourism Management, 40*, 382–393.
Cohen, E. H., (2011). Educational dark tourism at an in popular site: The Holocaust Museum in Jerusalem. *Annals of Tourism Research, 38*(1), 193–209.
Coldwell, W., (2017). *First Venice and Barcelona: Now Anti-tourism Marches Spread Across Europe*. Available from: http://www.theguardian.com (accessed on 20 January 2022).
Doyle, T., & Chaturvedi, S., (2010). Climate territories: A global soul for the global South? *Geopolitics, 15*(3), 516–535.
Freire-Medeiros, B., (2014). *Touring Poverty*. Routledge: Routledge.
Giddens, A., (2009). *The Politics of Climate Change*. Cambridge: Polity Press.
Giddens, A., (2011). *Runaway World*. London: Profile books.
Gotham, K. F., (2008). From 9/11 to 8/29: Post-disaster recovery and rebuilding in New York and New Orleans. *Social Forces, 87*(2), 1039–1062.
Gu, H., & Wall, G., (2006). SARS in China: Tourism impacts and market rejuvenation. *Tourism Analysis, 11*(6), 367–379.
Haigh, R., & Amaratunga, D., (2010). An integrative review of the built environment discipline's role in the development of society's resilience to disasters. *International Journal of Disaster Resilience in the Built Environment, 1*(1), 11–24.
Hajibaba, H., Gretzel, U., Leisch, F., & Dolnicar, S., (2015). Crisis-resistant tourists. *Annals of Tourism Research, 53*, 46–60.

Hammond, M., & Wellington, J., (2013). *Research Methods: The Key Concepts*. London & New York: Routledge.

Hartmann, R., (2014). Dark tourism, Thanatourism, and dissonance in heritage tourism management: new directions in contemporary tourism research. *Journal of Heritage Tourism, 9*(2), 166–182.

Holder, J., (1989). Tourism and the future of Caribbean handicraft. *Tourism Management, 10*(4), 310–314.

Hubner, A., & Gossling, S., (2012). Tourist perceptions of extreme weather events in Martinique. *Journal of Destination Marketing & Management, 1*, 47–55.

Johnston, T., (2013). Mark twain and the innocents abroad: Illuminating the tourist gaze on death. *International Journal of Culture and Hospitality Research, 7*(3), 199–213.

Keane, S., (2006). *Disaster Movies: The Cinema of Catastrophe* (Vol. 6). New York: Wallflower Press.

Kladou, S., Kavaratzis, M., Rigopoulou, I., & Salonika, E., (2016). The role of brand elements in destination branding. *Journal of Destination Marketing & Management*. doi: 10.1016/j.jdmm.2016.06.011.

Klein, N., (2007). *The Shock Doctrine: The Rise of Disaster Capitalism*. New York: Macmillan.

Korstanje, M. E., & George, B., (2021). *Mobility and Globalization in the Aftermath of COVID19: Emerging New Geographies in a Locked World.* Basingstoke: Palgrave Macmillan.

Korstanje, M. E., & George, B., (2022). *The Nature and Future of Tourism: A Post COVID-19 Context.* Palm Bay, FL: Apple Academic Press.

Korstanje, M. E., & Ivanov, S. H., (2012). Tourism as a form of new psychological resilience: The inception of dark tourism. *Culture – Magazine of Culture and Tourism, 6*(4), 56–71.

Korstanje, M. E., (2013). Preemption and terrorism. When the future governs. *Cultura, 10*(1), 167–184.

Korstanje, M. E., (2016). *The Rise of Thana Capitalism and Tourism.* Abingdon: Routledge.

Korstanje, M., (2014). Chile helps Chile: Exploring the effects of earthquake Chile 2010. *International Journal of Disaster Resilience in the Built Environment, 5*(4), 380–390.

Lacy, M. J., (2001). Cinema and ecopolitics: Existence in the Jurassic Park. *Millennium-Journal of International Studies, 30*(3), 635–645.

Ladkin, A., Fyall, A., Fletcher, J., & Shipway, R., (2008). London tourism: A 'post-disaster' marketing response. *Journal of Travel & Tourism Marketing, 23*(2–4), 95–111.

Leadbeater, C., (2017). Anti-tourism protesters in Barcelona slash tyres on sightseeing buses and rental bikes. Available from: http://www.telegraph.co.uk (accessed on 20 January 2022).

Lovelock, B., & Lovelock, K. M., (2013). *The Ethics of Tourism: Critical and Applied Perspectives*. London & New York: Routledge.

Mair, J., Ritchie, B. W., & Walters, G., (2016). Towards a research agenda for post-disaster and post-crisis recovery strategies for tourist destinations: A narrative review. *Current Issues in Tourism, 19*(1), 1–26.

Mihalache, M., & Mihalache, O. R., (2016). Organizational ambidexterity and sustained performance in the tourism industry. *Annals of Tourism Research, 56*, 128–163.

Park, C. W., Eisingerich, A. B., Pol, G., & Park, J. W., (2013). The role of brand logos in firms' performance. *Journal of Business Research, 66*, 180–187.
Park, S. Y., & Petrick, J. F., (2006). Destinations' perspectives of branding. *Annals of Tourism Research, 33*(1), 262–265.
Peters, M., (1980). The potential of the less-developed Caribbean countries. *International Journal of Tourism Management, 1*(1), 13–21.
Poria, Y., (2007). Establishing cooperation between Israel and Poland to save Auschwitz concentration camp: Globalizing the responsibility for the massacre. *International Journal of Tourism Policy, 1*(1), 45–57.
Quarantelli, E. L., (2005). *What is a Disaster? A Dozen Perspectives on the Question*. New York: Routledge.
Raine, R., (2013). A dark tourism spectrum. *International Journal of Culture, Tourism and Hospitality Research, 7*(3), 242–256.
Sanchez, P. M., & Adams, K. M., (2008). The Janus-faced character of tourism in Cuba. *Annals of Tourism Research, 35*(1), 27–46.
Sather-Wagstaff, J., (2016). *Heritage that Hurts: Tourists in the Memoryscapes of September 11*. Abingdon: Routledge.
Séraphin, H., & Gowreesunkar, V., (2017). On the use of qualitative comparative analysis to identify the bright spots in dark tourism. In: Korstanje, M., (ed.), *Virtual Traumascapes and Exploring the Roots of Dark Tourism* (pp. 67–83). Hershey, IGI Global.
Séraphin, H., (2014). *Tourism: Opening for the People of Toussaint?* Paris: Publibook.
Séraphin, H., Ambaye, M., Gowreesunkar, G. B., & Bonnardel, V., (2016). A marketing research tool for destination management organizations' logo design. *Journal of Business Research, 69*(11), 5022–5027.
Séraphin, H., Butcher, J., & Korstanje, M., (2016). Challenging the negative images of Haiti at a pre-visit stage using visual online learning materials. *Journal of Policy Research in Tourism, Leisure and Events, 9*(2), 169–181.
Séraphin, H., Gowreesunkar, V., & Ambaye, M., (2016). The Blakeley model applied to improving a tourist destination: The case of Haiti. *Journal of Destination Marketing Management.* doi: 10.1016/j.jdmm.2016.07.004.
Shondell, M. D., (2008). Disaster tourism and disaster landscape attractions after Hurricane Katrina: An auto-ethnographic journey. *International Journal of Culture, Tourism and Hospitality Research, 2*(2), 115–131.
Shondell, M., Gonzalez, C., & Hutter, M., (2017). Phoenix tourism within dark tourism: Rebirth, rebuilding and rebranding of tourist destinations following disasters. *Worldwide Hospitality and Tourism Themes, 9*(2), 196–215.
Smith, S., (2017). Organizational ambidexterity: Welcome to paradox city. *Human Resources Management International Digest, 25*(1), 1–3.
Stone, P. R., (2011). Dark tourism: Towards a new post-disciplinary research agenda. *International Journal of Tourism Anthropology, 1*(3, 4), 318–332.
Stone, P. R., (2012). Dark tourism and significant other death: Towards a model of mortality mediation. *Annals of Tourism Research, 39*(3), 1565–1587.
Tapper, J., (2017). *As Touting for Punt Trips Becomes a Crime, is Tourism Overwhelming Britain's Cities.* Available from: http://www.theguardian.com (accessed on 20 January 2022).

Tzanelli, R., & Korstanje, M. E., (2016). Tourism in the European economic crisis: Mediatized worldmaking and new tourist imaginaries in Greece. *Tourist Studies, 16*(3), 296–314.

Tzanelli, R., (2016). *Thanatourism and Cinematic Representations of Risk: Screening the End of Tourism* (Vol. 176). Abingdon: Routledge.

Urry, J., (2015). Climate change and society. In: *Why the Social Sciences Matter* (pp. 45–59). Basingstoke: Palgrave Macmillan UK.

Vitic, A., & Ringer, G., (2008). Branding post-conflict destinations. *Journal of Travel & Tourism Marketing, 23*(2–4), 127–137.

Walters, G., & Mair, J., (2012). The effectiveness of post-disaster recovery marketing messages—The case of the 2009 Australian bushfires. *Journal of Travel & Tourism Marketing, 29*(1), 87–103.

White, L., & Frew, E., (2013). *Dark Tourism: Place and Identity: Managing and Interpreting Dark Places.* London: Routledge.

Wilder-Smith, A., (2006). Tourism and SARS. In: *Tourism in Turbulent Times* (pp. 77–86). London: Routledge.

Wong, P. T., (2014). Viktor Frankl's meaning-seeking model and positive psychology. In: *Meaning in Positive and Existential Psychology* (pp. 149–184). New York: Springer.

Yang, Y., Zhang, H., & Chen, X., (2020). Coronavirus pandemic and tourism: Dynamic stochastic general equilibrium modeling of infectious disease outbreak. *Annals of Tourism Research, 83*, 102913.

Zhang, H., Wu, Y., & Buhalis, D., (2017). A model of perceived image, memorable tourism experiences and revisit intention. *Journal of Destination Marketing & Management*. doi: 10.1016/j.dmm.2017.06.004.

CHAPTER 12

Wellness Education and Practices After COVID-19: Holistic Approaches for Well-Being and Mindfulness

MARIA ALEJANDRA ZUCCOLI and MAXIMILIANO E. KORSTANJE

Department of Economics, University of Palermo, Larrea ST 1079, Buenos Aires, Argentina

ABSTRACT

In the constellations of tourism, many pre-graduate students simply fail to finish their careers or simply leave the university because of high competence and an emerging sentiment of psychological frustration. What is more important, once earned their degree the young professionals are subject to excessive working hours associated with low-paid salaries. There is a gap between students' ideals and the reality of the marketplace in the West. In Asia, a similar situation is claimed in a whole portion of literature. With basis on the importance of pleasure (wellness) to stimulate the student cognitive attention, PANCOE which is based on the Joy Labs (laboratory of Joy) at the University of Palermo, Argentina combines different techniques oriented to reduce the student dropouts as well as improving academic performance. Having said this, PANCOE centers efforts on the role played by pleasure and wellness in the education process. Interesting lines of research can be applied – following this methodology—in Latin American universities. PANCO transforms a climate of extreme competence which is often a source of psychological distress and frustration, in valuable cooperation (gamification) crystallized in a

Post-Disaster and Post-Conflict Tourism (An Updated 2nd Edition), Maximiliano E. Korstanje, Hugues Seraphin, and Vanessa G. B. Gowreesunkar (Eds.)

constructive cooperation process. The figure of play is vital to understand the PANCOE's success. PANCOE was applied to foreign students coming from neighboring countries such as Paraguay, Chile, Brazil, Bolivia, Ecuador, and Venezuela.

12.1 INTRODUCTION

In the line of a new century, new global hazards threaten the tourism industry. Experts agree that the industry is facing new radical shifts which demand innovative techniques in tourism education. The devastating effects of climate change (Airey & Tribe, 2006), as well as the claims of the local population for fairer forms of wealth distribution, without mentioning the problems of over-tourism and tourist-phobia, set the agenda of policymakers for the future (Cooper & Shepherd, 1997). Although paradoxically tourism education has notably increased worldwide the curricula were not updated according to the real problems of the industry. Many young students visit universities unfamiliar with the hard-working conditions of the industry and the sub-service sectors. This engenders a sentiment of discomfort, mistrust, and distress that elevate the rate of dropouts. Still furthermore, classic tourism education is certainly based on a much deeper Cartesian dualism which subordinates emotions to the power of the mind (rationality). The present chapter centers on the experience of PANCOE (an experiment conducted by the Joy Laboratory at the University of Palermo, Argentina). The experiment is ultimately applied to Latin American pre-graduate students who enter the first year of tourism Bachelor. Through the stimulation of pleasurable practices, the experience is originally designed to reduce the current drop-out rates as well as enhancing the students' academic performance. PANCOE combines different innovative techniques using digital technologies and the theory of emotions as the main theoretical framework. Although the obtained results are promising they cannot be extrapolated to other universes and contexts simply because PANCOE follows a non-statistical representative sample. The first section discusses critically the problems of tourism education in Asia. The classic curricula are far from being updated to the real problems of the industry. At the same time, students are pressed to compete with peers and subject to standardized protocols of skill training. The second section focuses on the role of joy and wellness to potentiate academic performance. At a closer look, new times need new tactics that mitigate the negative effects of climate change and the current economic downturn, without mentioning the aftermaths of COVID-19. Ultimately, details revolving around PANCOE are in depth debated in a third section. The future guidelines for research in Latin American countries are detailed in the last part of this chapter.

12.2 GLOBAL PROBLEMS OF TOURISM EDUCATION

One of the main problems of tourism education is mainly associated with what experts dubbed as "Cartesian dualism" which means that the rational mind subordinates the autonomy of the body. To put things in straight, the Cartesian dualism opens a gap between emotionality (and wellness) and rationality (knowledge production) (Rozemond, 1988). In tourism education, the Cartesian dualism not only dominates what is being published in leading journals but monopolizes the curricula content (Sheldon, Fesenmaier, & Tribe, 2011). Interesting new techniques of education as long-life learning are simply overlooked or pushed towards a marginal position (Su, 2014). Of course, Asian countries experience the same problems of tourism education as West (Zhao, 1991; Chaisawat, 2006; Du, 2003). In mainland China, the reforms initiated in higher tourism education for Asian countries have resulted in a difficult position for professors. Whilst the student's quota has been notably increased, creating in this way the rapid development of the industry, no less true seems to be that they accelerated long-dormant problems. Although tourism degrees and careers multiplied, the curricula are monopolized by the Asian governments following strong autocratic methods (Zhao, 1991). At a closer look, China lacks an international service ethic standardized in the international protocols. Besides, the explosion of higher tourism institutions is not enough to educate post-graduate students to solve the problems of the Chinese tourism industry. Post-graduate students look to get rapid promotions to be situated in managerial levels in a couple of years. Students are subject to great distress which is not comparable with their wages expectations. The syllabuses in China are designed to stress often on practical training misjudging the importance of skill learning to solve the problems of the industry (Zhang & Fam, 2008). What seems to be more important, new global risks such as terrorism, climate change or even the

COVID-19 pandemic threaten the future of the tourism industry. Education offers policymakers a fertile ground to follow with a creative solution to overcome the obstacles the industry daily faces (Farmaki, 2018; Fu et al., 2019; Hsu, 2005). No less true is that tourism education has recently been located as a promising area of study in tourism research above all in the context of emergency or post-disaster recovery. Tourism education should be defined as any process that trains the next policymakers, staff, researchers or even professionals who are legally designed to resolve the real problems of the industry to create a climate of synergy among stakeholders (Cooper, 2002; Belhassen & Caton, 2011; Paris, 2011). It is noteworthy that digital technologies support the formation of the next tourism professionals alternating creativity and entrepreneurship (Buhalis & Law, 2008). Over the recent years, some critical studies emphasize the dark side of technology alerting to its limitations to forge critical thinking. Digital technologies produce a great flow of information leading students to dispersion or simply to saturation. Whilst technology accelerates the decision-making process in some conditions, it reduces critical thinking ushering students into mechanical reasoning (William & Hobson, 1995; Munar & Gimothy, 2013). In this respect, one of the paradoxical situations of tourism education rests on despite the abundance of publications, little is known about its dynamic for the years to come (Sigala, 2020; Carr, 2020). Other scholars suggest overtly that tourism education entered an unparalleled state of crisis (Korstanje & George, 2021). It is difficult to resist the impression that the current tourism education crisis is given by a combination of factors which include the obsession for tourist-centricity and the economic-based paradigm (Franklin, 2014), the standardization of syllabuses to educate professionals not researchers (Tribe, 2002; Cuffy, Tribe, & Airey, 2012), the lack of coherence to offer all-encompassing models in tourism education given the multiple segments of the industry (Amoah & Baum, 1997), as well as the indifference of educator for sustainability issues (Gupta, 2016) without mentioning the crisis of sense of students who dream to get a job in a difficult marketplace mainly marked by low-paid wages and labor exploitation (Daniel et al., 2017).

Last but not least, the current curricula and syllabuses undermine the importance of joy and wellness to maximize high-quality experiences in education. Given the current academic formation, postgraduate students are unable to successfully resolve the problems of the tourism industry. Besides, workers are pressed to compete with their peers in quest of further profits. Hence recently published studies revealed that tourism staff shows higher levels of frustration and distress at their workplaces (Ross, 2006; Ashtalkosi & Ashtalkoska, 2012; Tsaur & Lin, 2014).

12.3 THE POWER OF WELLNESS AND JOY

For some reason, which is very hard to precise here, wellness has been systematically ignored in tourism studies. The concept proposes a wide range of subtypes such a medicine, public health, evasion, pleasure only to name a few (Sheldon & Bushell, 2009). Recently, some voices have enthusiastically applauded the role played by wellness-centered tourism in the sublimation of psychological discontent. Having said this, tourism should be understood as an instrument for human health and wellness. The distressing conditions of work, associated with day-to-day burdens, speak to us of the need of adopting vacation travels for health and wellness enhancement (Lehto & Lehto, 2019). Although wellness tourism remains a captivating theme to be studied in the constellations of tourism research, probably considering wellness tourism as the most ancient form of tourism in the Roman Empire (Smith & Kelly, 2006), less attention was given to wellness applied to tourism education. Thai, Smith, & George (2021) hold that wellness should be incorporated to improve pre-graduate students' competencies in tourism curricula. Based on empirical research with a focus on the Delphi method conducted in wellness travel agents, authors say that wellness not only would optimize students' performance but also laying the foundations towards a new foundational curriculum for under-graduate (UG) students. The specialized literature associate's wellness tourism only to the tourists' experience (Luo et al., 2018) but it may be very well applied in other contexts. For example, in Asia interesting forms of wellness-related experiences which include Yoga meditation, wellness nutrition, as well as many other physical and psychological activities can be added to tourism education. Lastly, Prebezac, Schott, & Sheldon (2016) alert the importance to adopt new imperative changes in tourism curricula, above all because of the dramatic changes the industry is going through. At least, skills and knowledge in tourism students must be re-defined. For the next decades, students would look for jobs that do not exist in our present. What tourism scholars learnt seems to be obsolete for the years to come. What is more important, the global ecological crisis presses educators for developing more values-based educational programs. Emotions and psychology play a leading role in the configuration of the next 21st century tourism curricula.

12.4 IN THE NAME OF JOY: THE PANCOE EXPERIENCE

The case study is based on different digital platforms such as Twitter and Facebook alongside multiple combinations of learning techniques. It is worth reminding the 5 human senses: smell, taste, touch, sight, and hearing). Participants were exposed to a positive stimulus (smell and taste), whereas they were tested in some cognitive skills. At a first glimpse, their skills were notably potentiated through the stimulation of pleasure (joy) experiences in eating or smelling some dishes. The level of innovation in the fields of tourism education is such that, the research center received international recognitions, and was shortlisted for the Reimagine education awards QS Stars awards 2016 (Wharton College). It is important not to lose sight of the fact that the laboratory also involves fine cuisine, made by well-known chefs specialized in neurosciences and gamification. The laboratory is providing students with all the skills and knowledge they need for their future careers. The experiment consists of a combination of pleasurable and memorable experiences (stimulating through the senses as tasting, smelling, and the gaze) whilst students are encouraged to cook. These pleasurable moments release endorphins that put students in the right frame to improve not only their attention but also stimulating their memory – above all for future exams. When taking exams, they are scoring much higher than students not involved in the program.

The present case study is designed to stimulate the learning skills of year 1 UG students registered on a Bachelor in tourism. Participants involved in the experiment are overseas students coming from Latin American nations such as Venezuela, Brazil, Chile, and Ecuador. The case explores the role of pleasure (joy) as an optimizer of learning abilities and skills whilst offering an innovative technique in tourism education and marketing. Students went through different teambuilding activities like cooking, baking bread, and food tasting. Current curricula in traditional tourism education do not consider the role of pleasure in improving the learning experience. PANCOE is a successful program used by the Laboratory of Joy to boost students' performances PANCOE has proven potential for stimulating socially deprived students as well as students living in politically unstable and unsafe destinations.

The experiment was run using two different focus groups (10–20 participants in each group). The main goal of the laboratory was to associate the creativity of gifted students with artificial intelligence (AI) and digital social networks such as Twitter. Secondly, the Laboratory of Joy explores and acknowledges the importance of smart ideas generated through the formation of safe social networks. The students mingling allow rapid optimization of learning. This study coins the term collective intelligence to denote the performance of groups in exams and tests contexts.

The experiment was based on a sample of 870 participants who regularly use Twitter (Twitter-@holapancoe). The sample divides into active and passive participants. Whilst the former refers to tourism UG students who are experimenting PANCOE (40 students), the latter refers to students – coming from other universities or areas; but not involved in the experiment (830 participants). The cohort ranges from 18 to 25 years old. The sample was formed by tourism students taking a specific course named as integration and communication. PANCOE is looking at integrating the senses of students with their emotions, academic performance, and the digital platforms. PANCOE devotes efforts to transform negative feelings like fear into positive ones like joy.

The main learning objectives are as follows:

- To facilitate the inclusion of foreign students who are joining a course as year 1 students in tourism and hospitality careers;
- To optimize the student learning experience, alongside reducing the rate of students who drop out;
- To implement innovative methods to improve traditional tourism education, using tourism marketing case studies.

The set of experiments lasted 18 encounters of 90 minutes each. The participants were pre-graduate tourism students versed in tourism management and marketing. The present study explores the power of pleasure (joy) as a strategy to better the standard education system in Argentina. Whilst pleasure was overlooked as an instrument to provide feedback to students, PANCOE (and the Laboratory of Joy) goes in the opposite direction. The Laboratory of Joy stimulates students' skills and performance through the articulation of pleasurable experiences. The results notably evince how those students who participated in PANCOE are performing better than those who are not involved in the experiment. The endorphins liberated by positive interactive communication pave the way for the rise of pleasurable experiences which contributes to better academic performances. As a promising method, PANCOE was originally designed to support the learning process of foreign students when arriving in a new country. The main goal was mainly to retain the student by reducing academic dropout, in other words, reducing the rate of students failing to get qualifications.

The experiment is based on two focus groups mentored by Alejandra Zuccoli. The focus groups are described below. The experiment takes place during different days of the elective year whilst students bolster a fluid dialog with other students as long as the experience.

- Focus Group 1: Participants are tourism bachelor students from the University of Palermo or UP, Buenos Aires, Argentina. They are divided into two groups (20 participants each). Each group works in isolation at the first stage. It is important to note that group A involves directly with the PANCOE program whilst group B is enrolled on the traditional programs and syllabuses for their courses. At this preliminary stage, students (in group A) were encouraged to bake bread as well as cooking and sharing traditional dishes with their peers. Each participant created a Twitter account and was required to have 50–60 followers, to communicate and share information about their cooking recipes. The interaction was based on the importance to sell their products whilst gaining traction respecting other competitive products.
- Stage 2: At this stage, students are encouraged to share their emotions and feelings. They are encouraged to share their testimony (fears, problems, and hopes during the experiences). In addition, students are tested through standardized and regularly based written and verbal exams. This stage was 2 hours long. Results were released 2 weeks later.

PANCOE is the name of the method which is distinguished as the only one that deals with pleasure (joy) as a key factor of stimulation and knowledge optimization in the learning process. The method – which surfaces after diverse experiments—is administered by Alejandra Zuccoli who serves as Director of the Laboratory of Joy. The term PANCOE comes from Spanish PAN (Pensamiento automatico naturalizado: automatic and naturalized thinking) and COE (comunicacion basada en el estudiante

– communication based on students). To be more precise, the method connects directly with the endorphins of the brain exploring not only the students' skills but also their creativity. The experiment is organized by the Laboratory of Joy at the University of Palermo, Argentina. At the same time, PANCOE is an emerging and promising experiment to include foreign students coming from different countries. Besides, many of the students who enter the university are certainly educated following traditional methods. The Laboratory of Joy integrates pre-graduate students coming from different Latin American countries as well as social classes. Methodologically speaking, we use the terms joy and pleasure indistinctively. One operates in the field of endorphins which disposes the body to a pleasurable state whereas joy circumscribes to a durable state of the mind marked by happiness (well-being).

Each student not only developed a packaged product but developed skills to compete with others (getting further followers). The Twitter account was anonymous to meet a genuine dialog as well as dismantling the emotional barriers to bolster a more fluid communication. This virtual interaction engaged his-her performance with the academic courses. At the same time, students were bombarded with pictures of paradisiacal pictures and landscapes which invited them to travel mentally. Their bread was sold, advertised, and promoted using Twitter as well as Facebook. By a combination of tasting, smelling, and cooking students release higher levels of endorphins that dispose them to improve not only the attention but also the academic performance (stimulating memory). The laboratory toys with the belief that students are over-exposed to digital technologies like cellular phones that distract them in the classroom. Having said this, half of the group was poor grades in their respective courses. In addition, students come from different countries as well as cultural backgrounds, a fact which makes it very hard a synergy or integration in the classroom. Initially, Professor Zuccoli noted students expressed distressing and negative emotions as higher levels of anxiety and fear. This begged two significant questions: How can teachers replace digital technologies in the classroom for students not to be distracted? How can be a syllabus adapted to foreign students who come from different cultures?

Centered on the logic of tourism branding or marketing, PANCOE is a simple method equipped with scientific measurements which consist in accomplishing pleasurable tasks like eating, tasting or even cooking to improve academic performance. The experiment originally instrumentalizes the cognitive reserve of endorphins to engage poor-grade students but – sooner than later – the method was applied to other groups. To create safer social networks, the goal of PANCOE associates with the integration of creativity and AI and digital technologies. PANCOE struggles to transform negative emotions or experiences into joy (well-being). As long as the experiment, each participant was invited to imagine the best-preferred tourist destination sharing with other participants the culinary customs of their own country. Likewise, students were subject to see pictures containing islands, beaches, and paradisiacal landscapes. We use the term imagined landscape to create a climate of well-being. The experiment takes place in three different stages.

The first stage follows the logic of tourism branding or marketing. Each student involves in creating a cultural product, a dish, or a bread piece. Participants are encouraged to create a Twitter account to promote their products. In fluid interaction with others, participants exchange recipes whilst answering questions about how the product was ultimately elaborated. Meanwhile, participants are forced to compete with others to gain further followers. PANCOE is based on 18 encounters where students share their experiences with peers but also with well-renowned Argentinean chefs without mentioning students and lecturers from other faculties. Complementarily, in each encounter students, taste cultural dishes cooked by their classroom mates. This process is vital to enhance students' interaction, allowing a rapid optimization of acquired learning and skills. Besides, the experiment re-integrates multiple sensory dimensions of students in a friendly-marked climate of cooperation. Furthermore, in each encounter, students are invited to imagine a preferred destination to enjoy their holidays. Each piece of bread exhibits not only a student's identity but a cultural imprint that facilitates interacting with peers more positively. Each Twitter account is anonymous and virtually followed by almost 50 persons.

At the second stage, students proceeded to express their emotions about the resulting experience by writing. Although because of privacy and ethics, their names have been distorted in the present study case, participants were interviewed on several occasions or at least invited to tell their experiences. Maria (female, 22 years old) said:

> *"The experience was outstanding, thrilling, it was as to apply everything we learned in the classroom but the real life. I did not how to create a Twitter account, I learned to use Twitter to upload my bread as well*

as communicating my experiences with unknown others. I was shy to communicate my feelings, PANCOE was a great experience if you ask me."

Following this, Julieta (female, 20 years old) replies:

"To my end, the PANCOE associates directly to the communicative process; each student disseminates a product (a bread) to share a message with professors, students, and other people. This exhibits a circular model where all we are senders and receivers! The experience interrogated me emotionally situating as a real challenge, but it was indeed gratifying!"

The radical shift from traditional education to a more innovative approach occupies a central position in the Laboratory of Joy. As Rosa puts it:

"It was a strange experiment, innovative, different. Professor Alejandra put to use Twitter as an instrument of education and emancipation whilst our products were promoted to the public. Since the Twitter account was anonymous, it was free to express my feelings.

It was very interesting the previous moment to promote my slices of bread; I did not know who my followers were. My user-made public not only the loaves of bread but also fine gastronomy. Twitter serves as a fruitful and useful instrument to engage with others." In the same direction, we get with Rogelio (male, 20 years old) who overtly acknowledges that:

"PANCOE is great! I never used Twitter; it took some time to learn to move in twitter's world. Anyway, I created my avatar; I offered my products whilst interacting positively with other users. This is the type of experience I am willing to make again and again and again!"

The problem of self-esteem and peer valorization was present in PANCOE as well. Marcelo (male, 20 years old) eloquently claimed that:

"I had a serious problem talking with others. This is my Achilles tendon, to the limitations I had because I am not familiar with Twitter, one should add I am abroad out of my home. I learned to cook many dishes and I enjoyed getting new experiences during this experiment!

Finally, Natalia (female, 18 years old) accepts:

"PANCOE" is a new form of education simply because it put me in the other's lens. In PANCOE, you put yourself in the other place to improve the relations as well as gaining further understanding of the "Other." Through Twitter and PANCOE, I discovered my inner world."

The final stage (third stage) consists of regularly based verbal and written exams; each exam lasted approximately 2 hours. The results matched the expectations as the students involved in the program performed better than the others (the capacity of memory storage increased notably to 33% in students who took part in PANCOE in comparison to groups who did not participate in the experiment (20%). Besides the number of students who graduated was 40% amongst those involved in the program against 28% for the others.

Amongst the limitations of the study, there is no need to say, that further investigation is needed. In that direction, the study was based on middle-class foreign students, excluding Argentinean peers. In the same way, there is no evidence of results in lower classes or groups subject to durable psychological deprivations such as students in zones of war, social conflict as well as ongoing political instability or in devastated areas. Promising approaches can be done to marketing issues adapting PANCOE to future products. Lastly, PANCOE shows an interesting potential to mitigate the negative effects of COVID-19 on consumers. PANCOE remains an innovative method that credits the important role of pleasure (joy) in tourism education.

12.5 LIMITATIONS AND PERSPECTIVE OF PANCOE IN TOURISM EDUCATION

As stated, PANCOE proffers promising applications to improve the cognitive attention of deprived or frustrated students. Initially, PANCOE was designed to standardize either the performance of foreign students living alone far from their homes or to low-rated students. In this context, PANCOE was applied to foreign students entering the first year of Tourism BA. The outcome can be extrapolated to countries to solve the current problems in the curricula mentioned above in the earlier sections. PANCOE can stimulate pre-graduate students by introducing an eco-friendly program whilst training the next tourism workforce. The experiment works stimulating the adrenaline opening the doors to new innovative techniques of education. Hence, PANCOE can be applied in emergency or post-disaster contexts as well as a post-COVID-19 environment in neighboring nations.

12.6 CONCLUSION

After further debate, the present case report shows how pleasure can improve the student's academic performance. Whilst pleasure was systematically debunked from the classic education programs, PANCOE goes in the opposite direction. Centered on the Joy Labs, PANCOE introduces pleasurable experiences and activities to simulate students' skills. The results evince those students who participated actively from PANCOE has developed a state of mindfulness and wellness increasing their degrees in comparison to those who have not taken part in it. At the same time, the endorphin liberated by positive interactive communication mainly disposed of better academic performances. Originally PANCOE was designed to standardize the learning process of foreign (Latin American) students. The goal was chiefly oriented to retain the student reducing academic desertion, which means the rate of students who fail to earn a degree. The results showed that PANCOE not only improved the academic degrees of participants but facilitated their final graduation. Of course, more research should be focused on how PANCOE helps psychologically – disturbed or deprived students who live in zones of war, conflict, or disaster, as well as whether the same results apply for other groups. Last but not least, though the outcomes are certainly innovative and exhilarating, they cannot be extrapolated to other universes. PANCOE is not based on a statistical representative method. PANCOE is a successful experiment performed by the University of Palermo, Buenos Aires, Argentina.

KEYWORDS

- **artificial intelligence**
- **communication**
- **environmentalism**
- **global hazards**
- **integration**
- **tourism education**

REFERENCES

Airey, D., & Tribe, J., (2006). *An International Handbook of Tourism Education.* Abingdon, Routledge.

Amoah, V. A., & Baum, T., (1997). Tourism education: Policy versus practice. *International Journal of Contemporary Hospitality Management, 9*(1), 5–12.

Ashtalkosi, S., & Ashtalkoska, I., (2012). Frustration of working as a problem of human resources management. *UTMS Journal of Economics, 3*(2), 193–198.

Belhassen, Y., & Caton, K., (2011). On the need for critical pedagogy in tourism education. *Tourism Management, 32*(6), 1389–1396.

Buhalis, D., & Law, R., (2008). Progress in information technology and tourism management: 20 years on and 10 years after the internet—The state of eTourism research. *Tourism Management, 29*(4), 609–623.

Carr, A., (2020). COVID-19, indigenous peoples and tourism: A view from New Zealand. *Tourism Geographies, 22*(3), 491–502.

Chaisawat, M., (2006). Travel and tourism education in Thailand. *Journal of Teaching in Travel & Tourism, 5*(3), 197–224.

Cooper, C., & Shepherd, R., (1997). The relationship between tourism education and the tourism industry: Implications for tourism education. *Tourism Recreation Research, 22*(1), 34–47.

Cooper, C., (2002). Curriculum planning for tourism education: From theory to practice. *Journal of Teaching in Travel & Tourism, 2*(1), 19–39.

Cuffy, V., Tribe, J., & Airey, D., (2012). Lifelong learning for tourism. *Annals of Tourism Research, 39*(3), 1402–1424.

Daniel, A. D., Costa, R. A., Pita, M., & Costa, C., (2017). Tourism education: What about entrepreneurial skills? *Journal of Hospitality and Tourism Management, 30*, 65–72.

Du, J., (2003). Reforms and development of higher tourism education in China. *Journal of Teaching in Travel & Tourism, 3*(1), 103–113.

Farmaki, A., (2018). Tourism and hospitality internships: A prologue to career intentions? *Journal of Hospitality, Leisure, Sport & Tourism Education, 23*, 50–58.

Franklin, A., (2014). Tourist studies. In: *The Routledge Handbook of Mobilities* (pp. 94–104). Abingdon, Routledge.
Fu, H., Okumus, F., Wu, K., & Köseoglu, M. A., (2019). The entrepreneurship research in hospitality and tourism. *International Journal of Hospitality Management, 78*, 1–12.
Gupta, V., (2016). Indian reality tourism-a critical perspective. *Tourism and Hospitality Management, 22*(2), 111–133.
Hsu, C. H., (2005). *Global Tourism Higher Education: Past, Present, and Future*. London, Routledge.
Korstanje, M., & George, B., (2021). *The Nature and Future of Tourism: A Post COVID-19 Context*. Palm Bay, Apple Academic Press.
Lehto, X. Y., & Lehto, M. R., (2019). Vacation as a public health resource: Toward a wellness-centered tourism design approach. *Journal of Hospitality & Tourism Research, 43*(7), 935–960.
Luo, Y., Lanlung, C., Kim, E., Tang, L. R., & Song, S. M., (2018). Towards quality of life: The effects of the wellness tourism experience. *Journal of Travel & Tourism Marketing, 35*(4), 410–424.
Munar, A. M., & Gyimóthy, S., (2013). Critical digital tourism studies. In: *Tourism Social Media: Transformations in Identity, Community and Culture* (pp. 245–262). Emerald Group Publishing Limited.
Paris, C. M., (2011). Social constructivism and tourism education. *Journal of Hospitality, Leisure, Sports and Tourism Education (Pre-2012), 10*(2), 103.
Prebežac, D., Schott, C., & Sheldon, P., (2016). *The Tourism Education Futures Initiative: Activating Change in Tourism Education*. Abingdon, Routledge.
Ross, G. F., (2006). Tourism industry employee workstress—A present and future crisis. *Journal of Travel & Tourism Marketing, 19*(2, 3), 133–147.
Rozemond, M., (1998). *Descartes's Dualism*. Cambridge, Harvard University Press.
Sheldon, P. J., & Bushell, R., (2009). Introduction to wellness and tourism. *Wellness and Tourism: Mind, Body, Spirit, Place*, 3–18.
Sheldon, P. J., Fesenmaier, D. R., & Tribe, J., (2011). The tourism education futures initiative (TEFI): Activating change in tourism education. *Journal of Teaching in Travel & Tourism, 11*(1), 2–23.
Sigala, M., (2020). Tourism and COVID-19: Impacts and implications for advancing and resetting industry and research. *Journal of Business Research, 117*, 312–321.
Smith, M., & Kelly, C., (2006). Wellness tourism. *Tourism Recreation Research, 31*(1), 1–4.
Su, Y., (2014). Lifelong learning in tourism education. *The Routledge Handbook of Tourism and Hospitality Education*, 322–329.
Thal, K. I., Smith, S. L., & George, B., (2021). Wellness tourism competences for curriculum development: A Delphi study. *Journal of Teaching in Travel & Tourism*, 1–16.
Tribe, J., (2002). Research trends and imperatives in tourism education. *Acta Turistica*, 61–81.
Tsaur, S. H., & Lin, W. R., (2014). Hassles of tour leaders. *Tourism Management, 45*, 28–38.
Williams, P., & Hobson, J. P., (1995). Virtual reality and tourism: Fact or fantasy? *Tourism Management, 16*(6), 423–427.

Zhang, W., & Fan, X., (2006). Tourism higher education in China: Past and present, opportunities and challenges. *Journal of Teaching in Travel & Tourism, 5*(1, 2), 117–135.
Zhao, J. L., (1991). A current look at hospitality and tourism education in China's colleges and universities. *International Journal of Hospitality Management, 10*(4), 357–367.

CHAPTER 13

Disruptive Transitions During COVID-19: The Tourism Industry in Context

GUILLERMO GARCIA[1] and BABU GEORGE[2]

[1]*Universidad San Pablo, Guatemala*

[2]*Christian Brothers University, USA*

ABSTRACT

In the midst of the crisis generated by COVID-19, the present chapter explores the limitations and problems of applied research to help in the tourism industry recovery. Based on an economic perspective, interesting outcomes and conclusions can be drawn to help the tourism industry to be on its feet back again. Starting from the premise that the COVID-19 affects seriously to the industry, important guidelines to mitigate its final impacts are presented to readers. At the same time, we caution on the necessary financial reforms to achieve more sustainable forms of consumption and recycling in the industry. COVID-19 reminds two important things. On one hand, the crisis opens the doors to an opportunity to get eco-friendly forms of consumption. On another, the implementation of these policies should be carefully studied and regulated by specialists.

13.1 INTRODUCTION

The COVID-19 crisis affects many people around the world who ultimately move in a climate of extreme uncertainty and fear. Having said this, the COVID-19 pandemic has generated difficult trade-offs whilst posing serious challenges to the tourism industry (Loayza, 2020). Over the

Post-Disaster and Post-Conflict Tourism (An Updated 2nd Edition), Maximiliano E. Korstanje, Hugues Seraphin, and Vanessa G. B. Gowreesunkar (Eds.)

recent weeks, the COVID-19 became in a pandemic that affected nations coming from four corners of the world. To the border and airport closure, one might add the closure of schools and universities affecting more than 1 billion of students in the spring of 2020. By the end of 2020, the pandemic has mushroomed, infecting more than 50 million people worldwide. As a result of this not only the tourism industry but also the global trade fallen in a paralleled crisis deriving in more than 1.25 million deaths (Hoseinpour et al., 2020). Governments desperately disposed of restrictive measures which included social distancing and lockdowns changing daily life at least as we know it (Bishop & Payne, 2021; Fenner & Cernev, 2021).

Estimates released by the OECD (Organization for Economic Co-operation and Development) for September 2020 reveals that the global GDP (gross domestic product) experiences a projected decline of 4.5 to rise up to 5% for 2021. According to OECD, the unemployment rate will be 9.4% in Q4 2020 in comparison to 5.4% in 2019. These forecasting suggests not only the virus outbreaks will keep on but also the mass vaccination is expected for the late of 2021 (OECD, 2020).

Tourism is a very high-touch industry (George & George, 2004; Gore et al., 2021); despite discussions around the future of virtual tourism, the element of real travel and physically visiting places cannot easily be replaced. The spirit of tourism is an experience in body, in a context disjointed from the place of residence of tourists (Hasana, Swain, & George, 2022). Given this, COVID-19 impacted a major jolt upon it (Fotiadis, Polyzos, & Huan, 2021). Primarily based on a review of literature, this chapter discusses the disruptive changes brought about by the pandemic in the special context of tourism.

13.2 CONCEPTUAL DEBATE

The long lasting and devastating effects of COVID-19 has been reported in many countries around the world, as stated with serious concerns that the lives of survivors, relatives, and friends are seriously harmed. Other causes of deaths associated to postponement of treatment because of the pandemic remain overlooked for specialists (George, Bucatariu, & Henthorne, 2020). Having said this, the number of bereaved people notably increased. Verdery et al. (2020) assessed the increasing number of people who have been affected by COVID-19 during the pandemic, estimating that the average number of those who experience a close loss. If 1,90,000 Americans died from COVID-19, then 1.7 million – or near circa—are bereaved. As noted, the social impact is immensely huge.

Unprecedented measures are followed to care the health communities whilst promoting safer care as long as the pandemic accelerated structural changes not only regarding the death rates but also in daily routines (Mirzaee, Jalalinejad, & George, 2021). As it is often mentioned, COVID-19 led strategies encompass decrees and bills by Governments and policy makers about follow-up during terminal illness, funeral-burial arrangements linked to physical distancing isolation as well as hygiene requirements.

There are important reasons to think that the above-described points engender sentiment of grief and grieving who have experienced the loss of a relative because of COVID-19 viral infection. This makes critical the experience in respect to other types of bereavement. To some extent, this body of specialized literature is overlooked by specialists. Hence, the main goal of this chapter associates to give readers a snapshot of the existent knowledge available about the experience of bereavement in the post-COVID-19 contexts whilst suggesting new approaches for applied research. What is more important, in early 2020, the economic impacts in the US situated as a slippery matter affecting every facet of the American economy. In the health industry, COVID-19 started with a paradoxical rise of resource utilization but decreasing in productivity and revenue (Sivakanthan et al., 2020).

13.3 THE TOURISM INDUSTRY CONTEXT

The tourism industry was least prepared for something like this (Korstanje & George, 2021). The situation was complicated all the more because tourism is not a simple homogenous industry at all. It is rather an emergent industry from the partial contributions made by various other industries; all these contributory industries had their own struggles, and the concerns of tourism were not always on the top list. For instance, airlines contribute significantly to international tourism; yet, as COVID-19 unfolded, it became apparent that airlines devoted considerable efforts in overcoming the barriers left by the pandemic. Some industrial sectors were more dependent upon tourists, say, cruise ships, amusement parks, gift shops, and restaurants in tourist destinations, among others. These were the ones most hit. Museums, zoos, etc., too were hit; but, given the way they are economically supported by governments and community organizations, they weathered the storm much better. These kinds of attractions also began to offer the virtualized offering of their attractions as a substitute, the euphoria did not translate to income earnings.

The experience of the tourism industry is a humble reminder that digitalization does not solve all the problems. There are aspects of human experiences as tourists that just cannot be virtualized without losing quality (Korstanje & George, 2017). That said, ancillary services that support such experiences could and should be transformed digitally. Building scalability into these services will ensure competitiveness. Platform models of product development and management will open up opportunities in this area to a large number of newcomers (Aghaei, Naderibeni, & Karimi, 2021). The pandemic led us to acknowledge the importance of the segment of domestic travelers who helped many a tourist business to stay afloat. In many destination areas, local communities were even looked down upon; with the decline in long-distance travel during COVID-19, the local community members, and those in neighboring communities at drivable distances began to make more and more visits to their local attractions. This should help change the viewpoints of the leaders of the tourism industry about the value of these communities. These communities are not merely a source of economic support but also a vital building block of social capital for tourism.

The tourist economy, the global system in general, is always in constant motion with or without COVID-19, which suggests that is not that something is right or wrong due to a pandemic; obviously a pandemic if it exerts influence on specific economic indicators, however, these can be positive or negative depending on the branch we refer to. The global economy was already shattered before the coronavirus (George & Rogers, 2018). Many of those on whom our survival depends on do not receive the corresponding payment. Caregivers, drivers, cleaners, nurses, supermarket cashiers keep on their works as stay-at-home advertising executives, marketers, stockbrokers, and soccer players. All of this is an indication that the world economy showed signs of sickness earlier than the COVID-19 hit. Fortunately, some remedies and cures are available. We can consume less and buy local whereas to regulate the extreme competence in the labor market as well as the working conditions of exploitations comes to an end.

Governments can pass bills and sanctions laws that need essential workers to be paid no less than the median wage in a country or more. In this token, we can reject to those leaders or fear-mongers who sow hatred and resentment and instead celebrate those who show compassion and sagacity (Castellanos & George, 2020). We can win back by the demand of further flexibility for workforce as well as abandoning stairs that are not worth climbing. We can stop worshiping wealth and start appreciating the arts. If we remember our environment before the COVID, we remember that the companies where we were working or with which we were related had the same economic, labor or success problems in the same areas, that means that if of course, the COVID brought us a change, but before that there were also good and bad things in the world.

13.4 TRANSITIONS DURING COVID-19

Walt Disney is an excellent example of an entertainment company that has made some profits but also losses during the pandemic. The company had to close its amusement parks when the closure measures were implemented. That has cost Disney at least $1.4 billion, according to CEO Bob Chapek. But at the same time, demand for Disney's streaming services has exploded. The Disney + platform, which launched in November, now has about 55 million subscribers.

The COVID-19 pandemic has wreaked havoc in corporate profits and business environments worldwide. The restrictive measures oriented to stop mobilities engendered many pitfalls and obstacles in the supply chain (Sharma et al., 2020) whilst placing the commercial activities into jeopardy (Rababah et al., 2020). These impacts involved not only small but great companies without mentioning medium-sized enterprises (Kalemli-Ozcan et al., 2020).

One of the main impacts left by COVID-19 seems to be directly on the supply side. In China, so to speak, factories systematically close ushering the economy into a standstill. The contraction of macroeconomic supply – which exchanges goods and services, is leading gradually to a stage of stagflation (Maital & Barzani, 2020). Most companies have had changes from the workplace: Home office, face-to-face, hybrid arrangements, etc. However, more than that is the businesses that have had to close globally or in specific countries due to not being able to market their products or services during the pandemic. COVID-19 has put the global economy on its knee but no less true is that lockdown measures restricting our movements have also helped some businesses prosper. Let's see some examples of what has happened to the economy, in his article "The Economy and the Coronavirus: The Winning Businesses and the Surprising Losers During the Pandemic," BBC News Mundo on May 14, 2020, gives us the following examples.

Curiosity and open-mindedness aside, let's look at another segment of the market that everyone knows, uses, but doesn't talk about sex workers. From Colombia to Denmark, there has been an increase in the sale of sex toys during confinement. It's a big business with a market that moved nearly $27 billion in 2019. COVID-19 seems to have given the sex toy industry a boost, with companies specializing in high-tech devices offering "long-distance experiences" that they benefit from social distancing. But the coronavirus has caused a loss of income and health risks for sex workers. In many countries, sex workers do not have legal rights and are not eligible for government assistance programs, leaving them poor and even homeless during the pandemic. Japan is an exception, as it is a country that has offered financial assistance to sex workers during this crisis.

There is a turning point in markets which mainly characterizes a rapid volatility in the financial flows. From January's prospects for an upturn in global growth records equate February's downgrades in the economic growth. This added assessment frames a certain type of heterogeneity in the COVID-19 effects which change from country to country (Mann, 2020).

If we analyze each business, it had things to form or against, however, some reinvented themselves and kept going despite changes, losses, and many changes. It is noteworthy that distance, social, and essential jobs have historically been less cyclical. This historical acyclicity of tele-workable occupations is attributable to their higher proportion of skilled workers. However, in opposition to the global stock and market crisis (2008), employment has abruptly declined while employment in essential and tele-workable jobs decreased lest as long as the pandemic crisis. In both recessions, anyway, low-paid workers have been vulnerated more than high-income workers, suggesting a significant distributional impact from the two recessions (Shibata, 2020).

13.5 THE FUTURE OF REFORMS POST-COVID-19

No one can definitively predict the future, especially after a pandemic like COVID-19, which is the first time we have suffered in the world. But for those who have lost a loved one, have lost their jobs, their lives have radically changed. As of June 2021, we cannot even talk about after the COVID since in many countries, not enough vaccines have arrived yet and many cases of contagion continue. However, we can share reports that make us think about what awaits us after the COVID, for example, the Forbes review reports in two articles.

The first one published on 10.15.2020 about the world's best employers, Forbes supported Market research firm Statista to compile fourth-annual list of the World's Best Employers. We made a survey over 1,60,000 full time and part time workers – who come from 58 countries. These surveys were made following a snowball method from June to July, where participants have been asked to rank their willingness to recommend their employers to peers and friends. They, also, were requested to give some rates for job satisfaction based on image, economic footprint, talent development, gender equality and social responsibility (only to name a few). The ultimate list is based on 750 multinational and large corporations located in 45 countries. Interesting article and it also shows us the changes that many companies have had to make to survive and give their employees a quality of working life as much as possible.

The second report is: How the World's Largest Public Companies, it endured the Pandemic of 12/05/2021, which indicates that the obtained outcomes are not all bad. Since central banks allow negative interest rates which associate with no alternatives to invest in stocks. Naturally, the total global stock market jumped about 48% last year. So, while sales and profits may decline for the Forbes Global 2000 companies, total assets and market value have increased (see our detailed methodology below, based on the latest 12-month financial data available to us on April 16, 2021). Sign of the Times: The minimum market value to consider for listing in 2021 was $8.26 billion, compared to $5.27 billion in 2020. Undoubtedly, the above makes us think that after the COVID that we hope will be soon, it will be a strong recovery of the economy at a global level because companies will always seek to recover from the current situation.

The expectations after the COVID are reserved derived from everything that happened during this time. We cannot ignore the realities that have happened in the world and in different corporations, for example, those related to the hotel industry, tourism, and airlines to name just a few, since they have had millions in losses. Nevertheless, a report published by World Bank dated 01/05/2021, the world economy is expected to increase by 4% in 2021, considering that that the initial distribution of vaccines against COVID-19 (coronavirus) expands to in the year. Anyway, the recovery – for some voices—is tended to be moderate lest lawmakers actively act to control the pandemic whereas implementing reforms that revitalize investment in view to the Jan 2021 edition of its World Economic Outlook Report under the auspices of the World Bank. Even if the global economy is growing after 2020 contraction, the pandemic has caused large numbers of illnesses and fatalities, leaving millions into poverty and psychological distress. The top short-term political priorities are controlling the spread of COVID-19 and ensuring the rapid and widespread distribution of vaccines. To support the economic recovery, the authorities must also facilitate a reinvestment cycle aimed at achieving sustainable growth that is less dependent on public debt.

While the world economy seems to enter into moderate recovery, policymakers face enormous challenges – in public health, budget policies all associated to the reforms that assure a robust growth, as David Malpas puts it. To overcome the negative effects of impacts of the pandemic and counteract the adverse factors affecting investments, a major push is needed to improve the business environment, increase the flexibility of the labor market and products, and strengthen transparency and governance.

As of July 2021, an estimated number of 3.3 billion persons have been inoculated globally, but this huge number includes a small number of countries. For many countries in the world, the vaccine has been denied. Although some organization struggles to distribute freely the vaccine, through COVAX and several donations, many underdeveloped countries not only had no access but also failed to achieve substantial levels of vaccination in the population. Inspecting on the track global vaccine access, we review several platforms to understand how the global equity really works.

Centered on the current vaccine rates, we also expect an increase so that the vaccine coverage goals are ultimately met (at least per the reports issued by WHO, IMF, or World Trade Organization). Per these estimations, 40% coverage by the end of 2021 and 60% in 2021 reach acceptable levels of inoculation.

13.6 DISCUSSION

We are prone to be infected with COVID-19 for factors outside of our control. Also, although vaccines are highly effective, there is still no 100% effectiveness We can find many results in which we really needed to have a big change around the world in our lives. Men and women were immersed in their jobs, all the time were running, they never had time for anything. In this Pandemic, we had the opportunity to make different activities, for example The use of masks, face masks, gloves, alcohol gel, measuring the temperature to enter any place, avoid the capacity in closed places, social distancing, long lines waiting to be attended in a bank, or in a store, because the service did not allow them to gather as much and the accesses became more restricted. Home Office, to be at home with our family more time, curfew, without vacation, no visits to the mall, no Disney, no Park, don't go to the movies, or another public place.

Nevertheless, we had more approaches as a family, table games, time to reed family book time, movie at home, like a relative we had an opportunity to know more our children, coexisting with all the family, for example, with our mother, our father, brothers, sisters, or other relatives who live in our home. Our families are also able to learn about the activities that those of us who go to the office carry out, who go to our daily jobs, since we had to continue working from home.

We had to immerse ourselves in the online classes of our children, we had to teach them what we did not know, and they had a very good attitude towards the pandemic since they took seriously the necessary care to survive, study, to adapt to technology, learn new programs, grow in knowledge to be able to continue with life even if it was different. In my neighborhood, for example, I could see how strong and compassionate young people made purchases for the elderly to prevent them from leaving home since they were the most likely to get infected. Every Saturday they spent asking if they needed something for their pantry and they made large purchases, long lines, and brought the necessary items for the neighbors who could not do the shopping themselves.

Recycling is a major aspect of health and hygiene and plastic recycling had attracted critical attention to the hospitality industry even before COVID-19 (Ramirez & George, 2019). Taking a reference from Sharma et al. (2020), with the fact that medical and health care waste (M&HCW) has situated as a main concern for authorities. In this vein, Sharma et al.

(2020) have alerted on the employing incineration as well as gas sterilization or carbonization to deal with contagion wastes. These protocols of safety and security are very hard to implement in lower-income countries or developing economies because of many reasons. These countries have higher barriers to access to technologies to make good scientist research at labs while they are slower in the proven efficiency in handling with bulk quantities of solid wastes (i.e., thousands of tons per day) with inconsistent physical composition, which is a very pertinent issue during COVID-19 pandemic (Sharma, 2020). Nevertheless, some middle or lower-income economies often valorize plastic waste produced from M&HCW as constructions of material for manufacturing civil engineering composites (viz., tiles, bricks, wood-plastic composites) as described by Goli et al. (2020) after suitable disinfection. It is worth mentioning that the high temperatures (110–250°C) and pressures (<180 MPa) involved during this process (Goli et al., 2020) would ensure complete sterilization of plastic waste and ensure virus free composites. Furthermore, Sharma et al. (2020) convincingly argues that the transition from M&HCW to the central facility needs more human intonement and capital investment. Other studies suggest that there is a risk of spread of disease because of the aerosol formation during transportation if some protocols are not sustained. What is more important, some authors appreciate the introduction of artificial intelligence (AI) to resolve the already mentioned problems as well as optimizing the collection and segregation of solid waste – even after COVID-19 (Vaveroka et al., 2020; Sharma, 2020). All this intelligence should be supported by the provision of a proper dataset (which is a task very difficult to accomplish). Digital transformation is a double-edged sword, especially if not done thoughtfully and intentionally (George & Paul, 2020). Even if we are sympathetic with new tips and suggestions in utilizing drones for the home supply of medicines and food (in a quarantine) no less true is that security agencies should become in a major handicap during the pandemic. The use of drones may be thwarted by the fluctuation of air density, weather conditions without mentioning mischief or theft (Torabbeigi et al., 2020). To adopt an efficient scheme in the pandemic, scientists should begin with explorations on traffic management system, standardization of the delivery method as well as other alternative methods. To cut the long story short, in view of the possibility of COVID-19 being air-born, the delivery package and drones should be earlier cleaned before and after the delivery to avoid future reinfections or the spreading of the virus. In the case of industrial companies, both the global financial crisis of 2008 and the COVID-19 pandemic impacted somewhat equivalently. Despite having produced much more profit between January and March, mega digital players like Amazon face rising costs to handle the surge in orders, which have forced it to hire 1,75,000 more workers. The company says it will have to spend $4 billion to deal with the spread of COVID-19, which includes providing its workers with personal protective equipment and conducting disinfection operations in its giant warehouses. That amount exceeds Amazon's earnings for the first quarter of 2019 ($2.5 billion). Amazon has long resisted recognizing unions, arguing that it prefers to speak directly to its employees about any concerns they have. Before it announced the cost of expenses derived from COVID-19, Amazon had received criticism for the treatment of its workforce during the pandemic for security reasons.

Although most industries in the US suffered from both events, the performance of most industries' stocks began to rebound following the quantitative easing announcements. Our results indicate that quantitative easing is effective in boosting investor confidence. We also find that the effect of quantitative easing in 2020 on equity performance is most significant for industries that are most affected by the pandemic (Chi & Wei, 2021). On the other hand, in the area of dentists, a model shows that the longer the COVID-19 mitigation/suppression measures are maintained, the greater the financial distress imposed on dental clinics, especially affecting those with higher operating costs (Schwendicke, Krois, & Gomez, 2020).

13.7 CONCLUSION

What could industries like tourism do in order to overcome the crisis? Given the numerous trial and error experiments that were undertaken over the last two years, it has now become apparent what works and what does not work. But a more important question is how to solidify these learnings and build them into the structures and cultures of organizations. Also, how could we build resilience not just inside our organizations but also into the superstructures supporting the tourism industry? In normal times, such investments to prevent the negative impact of a pandemic might very well be perceived as an extravagance. COVID-19 is one of those rare black swan events; once it is past, firms working with thin profit margins for survival might cut corners even more aggressively to remain competitive. Thus, the next question is how to be resilient, even assuming the aforesaid dynamic. We do not have answers but believe we are asking the right questions here. In the ultimate analysis, asking the right questions will at least help us anticipate alternative scenarios.

We cannot deny that we suffer many deaths worldwide, the suffering of families that lost one or more of their members; this pandemic shook our lives a lot. This pandemic made us more human, it reminded us that we can help each other, that we can give others a little of our time, that not everything in life is work, that at any moment we can lose our loved ones and we were not valuing them as we should do it (Korstanje & George, 2021). We had forgotten to thank and hope that a family member would come home healthy every day, that we could serve more, love more, that we could show the world that we can do anything. A pandemic had to come to remind us that we were humans, not working robots.

We remind ourselves that it truly means the value of life, even more so when it is the life of our loved ones. As human beings we return to our protective, loving nature, which wants to see a friend or relative leave a hospital, any stranger who is going through difficulties to survive, no matter if we know him or not, we just want to end with this nightmare that stole so many people close and dear, we want to go back to being the people who dedicate themselves to being happy, not surviving. As human beings, we have survived any challenge that life has given us, on many occasions we have been the creators of such situations that have endangered human life, animal life, and the conservation of the planet. We had to get to the point that due to chemical miscalculations, we created our own path to death.

We are creators, inventors, we have a high degree of survival, and we are always in search of new ideas, physical, chemical, engineering, etc. We only have to ensure that every act has a consequence, and we must not put at risk the only thing in the world that has no economic value, "Human life." Around the world, figures move daily, in which we see more people infected and more deaths due to COVID-19. It is something that we cannot change, but we can be more responsible to take care of ourselves and the people we love. We are the generation that managed to survive a pandemic that took a large number of human lives, but we are also the era that will make the necessary changes so that our new life leads us to be more careful, more conscious, more humane people.

KEYWORDS

- **administration**
- **artificial intelligence**
- **conceptual debate**
- **COVID-19**
- **finance**
- **global system**
- **innovation**
- **reforms**
- **socio-economic development**
- **tourism**

REFERENCES

Aghaei, H., Naderibeni, N., & Karimi, A., (2021). Designing a tourism business model on a blockchain platform. *Tourism Management Perspectives, 39*, 100845.

Bishop, M. L., & Payne, A., (2021). Steering towards globalization: Can a reformed G20 rise to the occasion? *Globalizations, 18*(1), 120–140.

Castellanos, J. D., & George, B., (2020). Boardroom leadership: The board of directors is a source of strategic leadership. *Economics and Business Review, 6*(1), 103–119.

Chi Chen, H., & Wei, Y. C., (2021). The global financial crisis and COVID-19: Industrial reactions. *Finance Research Letters, 42*, 101940.

Fenner, R., & Cernev, T., (2021). The implications of the COVID-19 pandemic for delivering the sustainable development goals. *Futures, 128*, 102726.

Fotiadis, A., Polyzos, S., & Huan, T. C. T., (2021). The good, the bad and the ugly on COVID-19 tourism recovery. *Annals of Tourism Research, 87*, 103117.

George, B., & George, B., (2004). Past visits and the intention to revisit a destination: Place attachment as the mediator and novelty-seeking as the moderator. *Journal of Tourism Studies, 15*(2), 51–66.

George, B., & Paul, J., (2020). *Digital Transformation in Business and Society*. New York, NY: Springer International Publishing.

George, B., & Rogers, J., (2018). *The Changing Landscape of Global Businesses: Principles and Practices*. Cambridge Scholars Publishing.

George, B., Bucatariu, L., & Henthorne, T. L., (2020). Technology in medicine: COVID-19 and the "coming of age" of telehealth. In: *International Case Studies in the Management of Disasters*. Emerald Publishing Limited.

Goli, V. S. N. S., Mohammad, A., & Singh, D. N., (2020). Application of municipal plastic waste as a manmade neo-construction material: Issues & way forward. *Resources, Conservation and Recycling, 161*, 105008.

Gore, S., Borde, N., Hegde, D. P., & George, B., (2021). Empirically mapping the evolutionary phases of tourism area life cycle (TALC): The case of Goa, India. *Tourism: An International Interdisciplinary Journal, 69*(3), 346–366.

Hasana, U., Swain, S., & George, B., (2022). Community-centered management of ecological resources for sustainable tourism: Benefits and costs. *International Journal of Professional Business Review: Int. J. Prof. Bus. Rev., 7*(1), 4–16.

Hoseinpour, D., A., Alizadeh, M., Derakhshan, P., Babazadeh, P., & Jahandideh, A., (2020). Understanding epidemic data and statistics: A case study of COVID-19. *Journal of Medical Virology, 92*(7), 868–882.

Korstanje, M. E., & George, B., (2021). COVID-19, terrorism, and conspiracy plot: Everyone is guilty until proven innocent. In: *Mobility and Globalization in the Aftermath of COVID-19* (pp. 135–151). Palgrave Macmillan, Cham.

Korstanje, M., & George, B., (2017). *Virtual Trauma Shapes and Exploring the Roots of Dark Tourism*. IGI global.

Loayza, N., (2020). *Costs and Trade-Offs in the Fight Against the COVID-19 Pandemic: A Developing Country Perspective*. World Bank Research and Policy Briefs, (148535).

Maital, S., & Barzani, E., (2020). *The Global Economic Impact of COVID-19*. Samuel Neaman for Nation Policy Research.

Mann, C. L., (2020). *Economics in the Time of COVID-19*. London: CEPR Press.

Mirzaee, S., Jalalinejad, R., & George, B., (2021). Country of origin, COVID-19 vaccine and the future of travel: A preliminary study in Iran. *Brazilian Annals of Tourism Studies–ABET*.

OECD, (2020). https://www.oecd-ilibrary.org/social-issues-migration-health/the-territorial-impact-of-covid-19-managing-the-crisis-and-recovery-across-levels-of-government_a2c6abaf-en (accessed on 20 January 2022).

Ramirez, A., & George, B., (2019). Plastic recycling and waste reduction in the hospitality industry: Current challenges and some potential solutions. *Economics, Management and Sustainability, 4*(1), 18–24.

Schwendicke, F., Krois, J., & Gomez, J., (2021). Impact of SARS-CoV2 (COVID-19) on dental practices: Economic analysis. *Journal of Dentistry, 99*, 103387.

Sharma, H. B., Vanapalli, K. R., Cheela, V. S., Ranjan, V. P., Jaglan, A. K., Dubey, B., & Bhattacharya, J., (2020). Challenges, opportunities, and innovations for effective solid waste management during and post COVID-19 pandemic. *Resources, Conservation and Recycling, 162*, 105052.

Shibata, I., (2020). The distributional impact of recessions: The global financial crisis and the COVID-19 pandemic recession. *Journal of Economics and Business, 115*, 105971.

Sivakanthan, S., Pan, J., Kim, L., Ellenbogen, R., & Saigal, R., (2020). Economic impact of COVID-19 on a high-volume academic neurosurgical practice. *World Neurosurgery, 143*, e561–e566.

Torabbeigi, M., Lim, G. J., & Kim, S. J., (2020). Drone delivery scheduling optimization considering payload-induced battery consumption rates. *Journal of Intelligent & Robotic Systems, 97*(3), 471–487.

Vaverková, M. D., Paleologos, E. K., Dominijanni, A., Koda, E., Tang, C. S., Małgorzata, W., & Singh, D. N., (2020). Municipal solid waste management under COVID-19: Challenges and recommendations. *Environmental Geotechnics, 8*(3), 217–232.

Ye, Z., Yang, J., Zhong, N., Tu, X., Jia, J., & Wang, J., (2020). Tackling environmental challenges in pollution controls using artificial intelligence: A review. *Science of the Total Environment, 699*, 134279.

CHAPTER 14

Sustainable Supply Chain Management in Tour Operations: A Tool for Traveling with Confidence in Troubled Times

KAREN CRIPPS

Department of Responsible Management and Leadership,
The University of Winchester Business School, Winchester, Hampshire, SO22 4NR, UK

ABSTRACT

The purpose of this study is to present an analysis of sustainable supply chain management (SSCM) in tour operations. This is an important sector for analysis in responding to the challenges and opportunities for many diverse players in 21st-century tourism. Through a qualitative, exploratory approach, a framework for SSCM actions is provided which draws on literature from tourism, supply chain and purchasing, iteratively informed through interviews with UK tour operators. Findings: The framework conceptualizes SSCM activities in four progressive, inter-connected categories: awareness, incentive, investment, and enforcement. Theoretical perspectives of paradox theory and shared value analysis are proposed to further information understanding of SSCM knowledge and practice. Theoretical perspectives of paradox theory and shared value analysis provide an original addition to the analysis of SSCM in tour operations. These are timely and well suited to building knowledge and responses to the 'troubled times' facing 21st-century tourism.

Post-Disaster and Post-Conflict Tourism (An Updated 2nd Edition), Maximiliano E. Korstanje, Hugues Seraphin, and Vanessa G. B. Gowreesunkar (Eds.)

14.1 INTRODUCTION

Recovery of the global tourism industry requires "longer-term transformation to more resilient tourism destinations" (Gössling et al., 2021, p. 15). After decades of discussion on 'sustainable tourism, the global pandemic has brought added impetus to the need for transformational change through tourism systems. The World Travel and Tourism Council's (WTTC) (2021) 'Lessons learnt' report identifies sustainability as a 'collective responsibility. This applies especially to tour operator supply chains which are highly diverse and fragmented, with package components including accommodation, transport, excursions, destination ground handlers and a diverse range of other supporting businesses such as food and craft. sustainable supply chain management (SSCM) refers to the management of environmental, social, and economic impacts of business activities through the supply chain. As intermediaries in the tourism system, tour operators have the potential influence to positively contribute to destination capacity building, while also supporting consumers to confidently make travel bookings at a time of uncertainty and change.

The challenges and benefits of tour operators engaging in sustainable business strategies have been shown across international studies (Goffi et al., 2018; Hamid et al., 2021; Khairat & Maher, 2012; Lozano et al., 2016; Tay et al., 2016; Zapata et al., 2018), which underlines the case for developing an understanding of SSCM as part of a wider strategic approach to responsible management. While the literature on sustainable tourism spans decades, the tragic scale of the COVID-19 global pandemic has attracted acute reflection on more sustainable business models. This chapter examines SSCM in the context of UK tour operations although it applies to other global contexts in which the sector ranges from mass-market players (which typically offer mainstream sun, sand, and sea products in addition to specialist brand offerings) through to small and medium-sized specialist businesses. The WTTC's (2020, p. 4) outlook for 'Recovery and Beyond' of the tourism sector recognizes sustainability as a pillar for change in a world that "has been reinvigorated to tackle social, environmental, and institutional sustainability."

This "renewed interest" is noted by the UK trade group 'The Association of British Travel Agents (ABTA, 2020, p. 8) to reflect wider societal trends which influence consumer demands, such as climate change activism. Concurrent with heightened awareness of inequality through the #BlackLivesMatter movement, business responsibility is under more scrutiny than ever before. There are calls for transformational change in business operations, to "tackle unsustainability at its source rather than as an add-on to counteract negative outcomes of business" (Bocken et al., 2014, p. 44). Tour operator suppliers are the 'source' of what might be perceived to be the 'problem,' while also paradoxically, the 'solution' in achieving the United Nations Global Sustainable Development Goals (SDGs). This chapter puts forward paradox theory as an appropriate lens for the analysis of SSCM in tour operations and provides a framework by which SSCM actions can be identified. This framework can be applied to structure progressive development of the sector, and a basis for comparison of approaches across global contexts.

14.2 PARADOXES IN SUSTAINABLE SUPPLY CHAIN MANAGEMENT (SSCM)

The so-called "war over-tourism" (Yang et al., 2021) regarding calls for 'new' tourism versus 'back to normal' tourism, highlights the tensions between an urgent desire to re-build economies, alongside transforming systems to be more in alignment with the SDGs (Gössling et al., 2021). Building resilience in supply chains is a key factor in driving tourism change and recovery, whilst also heeding warnings such as Hall, Scott, & Gössling (2020) that recovery might only bring 'more of the same. Paradoxical tensions are at the heart of debates around sustainable development as an 'oxymoron,' or contradiction in terms. Within a business paradigm where the purpose of business is to "produce profitable solutions to the problems of people and planet" (Mayer & Roche, 2021, p. 11), then all actors in the supply chain need to simultaneously address commercial alongside sustainability goals. Paradox theory is well-suited to research into interdependent forces and actors as it offers a tool for transformational research and action into the 'new normal' (Sigala, 2020).

Paradox theory focuses on accepting, and indeed, embracing contradictions rather than attempting to resolve tensions (Smith, 2011). At the entry level, this requires supply chain actors to resist the dominant commercial logic to prioritize cost efficiencies (probably at the expense of one organization over another) but to adopt a more systemic, longer-term view. This calls for 'paradoxical thinking' which Hahn, Pinkse, Preuss, & Figge (2014) set out can be analyzed at multiple levels from the wider context, through to organizational systems and tensions faced at the individual level. Context is particularly critical in considering the situational contexts of globally diverse tour operator suppliers which are likely to hold differing cultural views of what is an acceptable or feasible business practice, what is required by regulation, differing economic priorities, and differing capacities. This is of note since earlier research identified that operators do not always take a long-term view of destination development and claim to lack control over impacts on destination (Carey et al., 1997; Curtin & Busby, 1999; Forsyth, 1995; Klemm & Parkinson, 2001; Miller, 2001; Tapper, 2001). Context also captures systemic power positions which influence the adoption of SSCM practices, which is especially notable in a sector where the purchasing power of operators in supplier relationships has been well documented (Bastakis et al., 2004; Buhalis, 2000; Medina-Muñoz et al., 2003; Tveteraas et al., 2014).

At the organizational level of paradox analysis, traditional purchasing functions might for example favor reliability, quality, and cost while sustainability standards favor environmental and social features. Through the process of contextualizing, it is proposed that managers can apply "paradoxical rather than adversarial sensemaking" (Xiao et al., 2019, p. 4) through a deeper understanding of the root causes of tensions, coming to accept and embrace them, rather than acting from a position of defensive posturing and avoidance. It is important to note that a paradox perspective does not imply that a company is abandoning a profit perspective but rather that it is looking at a wider range of sustainability issues than those that offer immediate business benefits (Hahn et al., 2017). It thus offers a tool for transformative action, addressing not only the limitations of business models based on shareholder/profit primacy, but one which facilitates consideration of all ecosystem players and impacts and can thereby be considered an approach for improved decision making.

14.3 SUSTAINABLE SUPPLY CHAIN MANAGEMENT (SSCM) – A TOOL FOR SHARED VALUE CREATION

SSCM offers a tool to develop communication and understanding between partners, which can potentially address the perceived lack of power of suppliers to influence the way they are marketed by tour operators (Richards & Font, 2019; Schwartz et al., 2008). Furthermore, it has been shown that engagement in environmental programs, certifications, and awards can enable suppliers to achieve a price premium (Picazo, 2018). Research has provided some evidence of operators beginning to engage in SSCM practices. This has contributed to the understanding of operator attitudes to managing different parts of the supply chain (Font et al., 2008), influences on environmental SCM approaches of large operators (Budeanu, 2009), processes to implement SSCM into operators' businesses (Schwartz et al., 2008), enablers of SSCM (Sigala, 2008), the importance of connecting quality, health and safety alongside sustainability in supplier auditing (Baddeley & Font, 2011) and the influence of different organizational structures in sourcing and selling sustainable packages (Richards & Font, 2019). While these studies provide good insights into organizational processes and barriers/enablers to implementation, less attention has been given to the specific activities that operators can implement with suppliers to engage in SSCM.

A recent study goes some way to fill this gap by classifying tour operator actions as 'unconnected' when low-cost and ad-hoc, as compared to 'high intensity' actions in which operators invest time and resource in the changes they want to see and sustain these over time (Bonilla-Priego et al., 2021). Their study suggests that a combination of both ethical and strategic intent is necessary for transformational sustainable change and to drive further capturing of value (Bocken et al., 2014) further research is needed into organizational and behavioral characteristics of business models based on high intensity and connected actions. This is important in analyzing SSCM from a perspective of 'shared value creation (Porter & Kramer, 2006). Font, English, Gkritzali, & Tian (2021, p. 2) contend that "sustainability can only be mainstreamed if the different stakeholders attach some cultural, social or emotional value to it." Aligned to this, is the recognition that all suppliers have the potential to 'create or destroy value' (Bocken et al., 2015). The following sections respond to this imperative and set out the design and principles of a framework that can be applied to the analysis of SSCM actions.

14.4 SSCM FRAMEWORK

An exploratory, framework-building research approach was adopted based on literature from tourism, supply chain management and purchasing. This was iteratively refined through purposive sampling and depth interviews with 42 leading UK industry players; 17 interviews with specialist operators and 25 interviews with managers across 3 mass-market operators. Although these were conducted pre-pandemic, the impact of the pandemic is considered throughout this chapter, and this is not considered to have influenced the analysis. As is symptomatic of qualitative research, the outcomes are "suggestive, rather than conclusive" (Crotty, 2006, p. 13).

However, through seeking feedback from participants on research findings following the interviews, and richly presenting the results through the use of participant quotations, the framework represents the development of informed, interpretive insights.

SSCM activities are shown in the framework (*see* Figure 14.1) to be actions or approaches that operators can adopt to address business impacts with destination suppliers. The framework conceptualizes SSCM activities in four progressive, inter-connected categories: awareness, incentive, investment, and enforcement. This reflects a progression in the relative resource intensity of each SSCM activity. Following 'awareness-raising' activities (dialog, guidance tools, assessment/feedback), it is proposed that tour operators implement 'recognition' activities (purchasing, promotion), to then possibly progress towards 'investment' (workshops, pre-payments/investments) and/or 'enforcement' activities (contractual criteria, purchasing sanctions). The connecting arrows indicate that this is a cyclical process and tour operators are likely to return to activities over time to address particular business impacts, suppliers, products or destinations.

FIGURE 14.1 A conceptual framework of SSCM activities.

The framework offers a diagnostic tool which conveys a sense of progression in SSCM engagement while recognizing inter-linkages and offering scope to analyze within the context of different operational and relationship environments. The following section sets the framework concepts, drawing on quotations from the interviews. These are labeled as (S) to indicate specialist operators, and (M) to indicate mass-market operators, followed by a letter that represents differing representatives interviewed within the same business.

14.5 AWARENESS ACTIVITIES

The framework proposes that operators begin to address SSCM through awareness-raising activities, focused on developing supplier understanding of operators' sustainability strategies through dialog, guidance tools and assessment/feedback. This is likely to take place concurrently with operators developing internal engagement.

14.5.1 DIALOG

Informal communication can engender supplier support and generate new ideas and expertise (Walter, 2003). It also provides purchasing personnel with the opportunity to learn from suppliers' localized knowledge (Modi & Mabert, 2007), can encourage suppliers to act when the benefits of investments are not immediately obvious (Hartley & Choi, 1996; Newman & Rhee, 1990), and build supplier commitment and motivation (Tan et al., 1998). These are important concepts when managing a range of different suppliers, in which management may not support investment in sustainability initiatives, and/or may lack knowledge in how to implement practices (Al-Aomar & Hussain, 2017). Communication is likely to be a particular challenge among larger businesses which are characteristically subject to pressure from tight overheads and restricted time available to spend with individual suppliers (Bastakis et al., 2004). This approach is therefore likely to be most appropriate when the number of suppliers used is limited, suppliers are highly valued, operators can build up in-depth knowledge of supplier businesses, or do not consider that written tools are an appropriate communication tool.

Informal dialog is a key feature of many specialists, linked to concepts of trust and commitment to sustainability as key indicators of supplier relationships, meaning there is less need for formalized sustainability criteria (as also found by Richards & Font, 2019). This is associated with the relatively small size of many operator and supplier businesses, and a preference to tailor communication to individual supplier characteristics.

Specialists may consider that dialog is so in-depth and informed by many years of doing business with the same suppliers, that it represents a style of consultation:

> *"In most of our destinations, we've worked with the suppliers for a very long time now. It is a sort of dialog, we talk about the staff, the management and various aspects of the way lodges are run" (S12).*

In this respect, Buhalis (2000, p. 118) finds that some specialists are "considered by hoteliers as partners and even friends" can be reversed to reflect operator sentiments.

14.5.2 GUIDANCE TOOLS

Guidance tools may include operator sustainability policies, codes of practice, manuals, leaflets, or posters, and maybe developed internally, or by external organizations. These can be useful to reach many suppliers but might present some challenges in developing common messages which are appropriate to a range of supplier characteristics. Alternatively, some operators may opt to direct suppliers to trade association networks or voluntary bodies, but this is dependent on their accessibility at a local level.

Specialists generally show a preference for more informal, dialog-based communication but there is increasing recognition of the value of guidelines to provide suppliers with a clearer indication of the nature of products the operator wants to buy, and/or to act as a timely reminder of desired practices. This was felt to be particularly useful where operators rely on ground agents to identify and manage suppliers. Overall, views vary depending on geographical and technical considerations such as ease of communication based upon technological infrastructure, language ability, and general understanding and attitudes towards sustainability.

Particular concern was expressed regarding the risk of offending suppliers that have more intimate knowledge of the local area:

> *"The suppliers we are dealing with are at the emerald end of being environmentally sustainable, and I can take you to a guy who will give you the Latin name of every tree in the park, what affects them, the ecology of the animals, how they work. He's acutely aware of the impacts that his lodge is having, much more than I am. And for me to send him*

out a couple of pages of guidelines, say to recycle his paper, when he's 250 km from the nearest place where he can buy an exercise book, is arrogant" (S17).

Operators of differing scales state a preference for a supportive approach in guidance documents:

"It is an aspirational document which is aimed at helping suppliers to improve their practices. It makes suggestions as to what they can do about issues, and if that's not feasible for them, it suggests alternatives" (M2a).

While this is a cost-effective method of sharing information, it is dependent upon suppliers to have the resources and capabilities to effectively respond. It is also important that any such tools are compiled with the input of sustainability, health, and safety and quality measures to ensure consistency of messaging (Baddeley & Font, 2011). There is also some risk that guidance tools such as codes of conduct are used to transfer responsibility to suppliers (Touboulic et al., 2014; Zapata et al., 2018; Richards & Font, 2019), without any meaningful contributions by operators.

14.5.3 ASSESSMENT AND FEEDBACK

Initially, sustainability assessments can serve as a useful knowledge-building tool regarding the nature of impacts across the business (and later they might be introduced as a purchasing requirement). They may be conducted through operators visits, possibly as part of health and safety assessments, self-completed by suppliers (ideally with a system of verification), or through a third party such as Travelife. The ABTA manages the 'Travelife' certification scheme to recognize its' members' use of accommodation suppliers which meet social, environmental, and economic impact standards. Such programs represent a key tool to understanding impacts and raising awareness:

"I think the first thing is to find out exactly what the issues are, we're not even at that point that we know what the impacts, we don't know what the issues are here, or in the resort. We don't know where we are, we are

> *right at the start of this. So that would be the first stage, these are the key issues, and then what can we do about any of those and formulate a strategy from there" (M1a).*

This could be positively received by suppliers, who might be very receptive to operator suggestions of how to implement business improvements. Conversely, some concern was noted in introducing more assessments, on the basis that operators perceive suppliers to already feel pressurized to meet health and safety standards which may exceed local regulatory requirements:

> *"We're talking about too much information, too much prying. Sometimes I think they feel that the tour operators can be quite intrusive, and that might be taking things too far" (M3b).*

Some specialists consider that regular and ongoing dialog is more suited to reflect the individual characteristics of suppliers, and suits the more informal nature of both businesses, in addition to perceived resource implications associated with 'paperwork:'

> *"This is all incredibly time-consuming; we are trying to make it less formal because if you get into assessments, you only get about 5% of our suppliers who will be interested" (S5).*

On the other hand, it was noted that more formal assessments could clarify company thinking, improve the efficiency of feedback processes, and emphasize the importance of sustainability issues to suppliers:

> *"I'm sure that formalizing things would enable us to have a greater understanding of what's going on and to help the suppliers to understand how serious we are about this" (S7).*

A formal assessment can represent an important SSCM activity since it identifies where improvement efforts should be focused (Hines, 1994), and provides a benchmark to evaluate the outcomes of activities (Prahinski & Benton, 2004). It can also provide the basis for winning management support for sustainability efforts and provides information that operators need to guide sustainability purchasing decisions. Since suppliers are often already subject to a range of health and safety and quality-related assessments, it is important that these are approached in a supportive manner, and that communication is synchronized across differing audits. After formalizing assessment and feedback processes, it is necessary to set out how this information will be used, and which actions will be taken to address areas of impact.

14.6 RECOGNITION ACTIVITIES

Either following on from or alongside awareness-raising activities, it is proposed that operators adopt recognition activities (promotional or purchasing incentives). The term 'recognition' is used here as it overcomes concerns by some operators that 'incentives' or 'rewards' imply some sense of operator superiority and fails to capture that some suppliers are pro-actively engaged in advanced forms of sustainability management.

14.6.1 PURCHASING INCENTIVES

Purchasing incentives may include preferred supplier status in system searches, longer-term/guaranteed contracts, and/or increased purchasing volumes. The informal nature of specialist operators generally precludes interest in recognition-based tools such as preferential status, longer-term contracting, or enhanced payments. For some, incentives such as better payment conditions are irrelevant, since this is considered part of standard business practice:

> *"I don't believe it should be bribery, I think that good practice justifies itself. We are encouraging our staff and suppliers to follow certain guidelines because we believe it's the right way for us to run our business or the right way for them to run theirs, but to say to someone 'if you do it we'll buy this,' I don't think it's the right way to do it" (S10).*

It was recognized that such an approach may be more resource-intensive than leveraging supplier performance through the threat of sanctions, but this is felt to be more suited to the managerial outlook in terms of creating high service levels and good working relationships:

> *"You can draw up policies with penalty clauses and all this and hammer guys into submission, and you probably make more money in the long run [...] but we're not here primarily in a race to get rich" (S9).*

The inclusion of sustainability criteria in purchasing decisions may require fundamental adaptations to organizational processes in larger operators, and it was, therefore, suggested that this could only be a gradual process:

> *"I think volume, time and time constraints of the purchaser because he will be in a resort contracting about twenty hotels in three days! It's fairly pressurized when you've got a certain amount of time in a resort. It's just a matter of education and building it in as we move forward" (M1e).*

Overall, it is likely that sustainability will only be a purchasing differentiator across mainstream operators when all other purchasing criteria are met.

14.6.2 PROMOTIONAL INCENTIVES

Unless sustainability is firmly integrated into business values and strategy, and possibly regulatory standards, it is more likely that operators try to engage suppliers through promotional opportunities such as classification schemes/sustainability management standards than formal integration into purchasing criteria. There are many differing awards and eco-labels used across the travel and tourism sector and it was felt that the number of programs available are confusing and are often not recognized by customers. This concurs with the earlier findings of Font, Sanabria, & Skinner (2003). While it is positive that the UK's trade association encourages members to engage in the Travelife accommodation certification system, visibility is restricted. As certification is limited both in terms of operator engagement and to principally only accommodation suppliers, it is apparent that the industry is still at an early stage of uptake in promotional certification approaches.

14.7 INVESTMENT ACTIVITIES

Following on from awareness and recognition-based activities, it is proposed that operators may adopt an investment approach to SSCM. This may include offering training opportunities to supplies or providing direct investments and/or pre-payments to suppliers. Such activities are only likely to be implemented on a targeted basis, among suppliers with high strategic value, and strong relationships.

14.7.1 WORKSHOP TRAINING

Training can be an important contributor to long-term capacity building and is a particular feature of operations that rely on tour guides' specialist knowledge and interpretation skills (Black et al., 2001; Curtin & Wilkes, 2005; Weiler & Ham, 2002). Operators with a high level of expertise in sustainability comment that this can be very detailed:

> *"Effectively we consult with them. They come to us saying 'this is what we want to do,' so it's essentially initiated by them, and then we work with them to select some land, and to set up the camp" (S12).*

However, this is a resource-intensive activity, and it is therefore unsurprising that as identified by Tapper (2001), operators commonly tend to pass on training responsibilities to ground agents. It is only therefore likely to feature among operators which place a high value on suppliers and in destinations where the required service level is lacking. It might be offered as part of health and safety training workshops, demonstrating the links between sustainability and quality management. Although there are some examples of joint training programs among larger operators which share suppliers, specialists note the limited scope for such engagement:

> *"One of the things that we try to do is secure our supply chain so that we can: (a) have more influence over it; and (b) we know that our competitors don't have access to it" (S10).*

This statement is reflective of the power and control dynamics at play in securing suppliers in competitive destinations, which might potentially imply sustainability is not a key requirement over quality and cost and limits collaborative action.

14.7.2 PRE-PAYMENTS/INVESTMENTS

It has been reported that where good quality accommodation is in short supply, operators have co-financed developments on a short-term basis (Humphreys et al., 2004; Quayle, 2000), and this could relate to SSCM if activities are linked to improved quality. Vertically integrated operators with accommodations and other suppliers within the group may have a particular opportunity to make such investments, associated with possibly higher levels of trust and commitment between partners. Smaller operators may also make such investments based on long term relationships with suppliers, especially where it improves product quality (such as more environmentally friendly land transport). Therefore, although investment approaches may seem rather aspirational, when tied into product and quality management they become more commercially realistic. Baddeley & Font's (2011) study identified that any environmental cost savings accrued by suppliers were unlikely to be re-invested in further sustainability actions. While paradoxical to not invest in further actions that bring sustainability benefits, faced with financial challenges, any actions that are directly linked to cost savings are likely to be limited.

14.8 ENFORCEMENT ACTIVITIES

Finally, it is proposed that operators may adopt an enforcement-based approach to SSCM which can include integrating sustainability into contractual criteria and/or applying purchasing sanctions. This is only likely to occur over the longer term, following awareness, recognition, and possibly, investment-based activities.

14.8.1 CONTRACTUAL CRITERIA

The integration of performance requirements into contracts is useful to reduce the possibility of opportunistic behavior, and to provide clarity among both parties of what is expected. Furthermore, fair purchasing and contracting procedures are fundamental to SSCM, which depend on stability in the operator-supplier relationship and are critically important in using bargaining power responsibly (Kalisch, 2002). This is a contentious issue since research has identified that operators do not always offer suppliers secure contracting conditions (Bastakis et al., 2004; Buhalis, 2000).

Many specialists do not form legally binding contracts with suppliers, which reflects the informal nature of the relationship, and that many businesses are of a manageable enough size to rely on building relationships based on mutual trust and commitment:

> *"We don't have formal contracts with any of the suppliers, it's more a looking after each other contract that we have" (S6).*

Specialists that do contract with suppliers perceive limited feasibility of integrating sustainability as a contractual condition, due to limited supply availability and supplier capacity. Some were wary of damaging working relationships through obliging suppliers to meet standards that may not fully consider local conditions:

> *"It's quite difficult I feel to impose it on them without really good, detailed discussion because we are taking people to their country, and we have to respect how they do things" (S11).*

Larger operators are wary of integrating clauses regarding sustainability, which are not subject to legal requirements, and would require significant adaptation to purchasing processes, alongside a monitoring and enforcement system:

> *"It's important to understand how it fits into a monitoring process. It's all very well saying it's in the policy, and it's in the terms of the contract, but if you don't have any way of enforcing it, then it becomes pointless" (M1j).*

It was highlighted that if enforcement of contractual criteria involves 'dropping' suppliers, this would not be commercially feasible if they met all other costs, service, and health and safety requirements:

> *"We can't afford to axe properties, and by building it into a contract, the implication is that if they are in breach of contract, then we are going to react to it in some way" (M2a).*

It was also expressed that operators should be able to demonstrate to suppliers that initiatives to improve performance are also being taken within head offices:

> *"I think we'd need to get a few 'wins' under our belts and a few practices established before we start building them into contracts. I feel a little uncomfortable asking suppliers to do something we don't practice ourselves" (M2a).*

Such a comment captures the tensions in SSCM in terms of driving forward commitment and practice among global suppliers that face differing destination level challenges while demonstrating at least equal performance across all other head office functions.

14.8.2 PURCHASING SANCTIONS

In some industrial contexts, buyers may impose sanctions for poor performance, for example by withdrawing business, applying competitive pressure through buying from multiple suppliers (Krause & Scannell, 2002), or basing purchasing decisions on assessment results (Sriram et al., 1992). However, it is unlikely that tour operators will adopt such approaches regarding sustainability performance, both because of the lack of regulatory stipulations, and the risk of limiting purchasing choice (especially where suppliers can easily switch to less stringent operators).

It was suggested that sanctioning agreements may be made based on health and safety sanctioning processes:

> *"The way we've done it about health and safety has been to work together on an improvement plan with suppliers within certain time*

schedules, and if within those times suppliers haven't complied, then we have taken properties out of the program, so that's an option as a way forward" (M3c).

Similarly, a larger specialist commented:

"We may have to give them a rap over the knuckles sometimes and say, 'you have to comply with this otherwise we may not work with you in the future" (S14).

These comments are indicative of the power relationship between operators and suppliers and could potentially negatively impact supplier businesses that are already pressurized to meet servicing quality requirements, at low cost.

There is evidence that operators have taken a coercive approach to influence supplier practices through collectively threatening to withdraw business if particular standards are not met:

"In a venue like that it's a lot more difficult to get them to do something off your own back than if you get all the other operators involved behind you, then they have to do it because otherwise, we stop sales, and their venue is empty" (M3f).

However, this comment refers to servicing characteristics, and it is likely to be some time before operators adopt the same approach towards sustainability, unless directly related to product quality and customer satisfaction.

14.9 CONCLUSION AND FURTHER RESEARCH

This study makes a specific contribution to tourism practice and knowledge in two ways. Firstly, it conceptualizes SSCM activities in a progressive categorization (awareness, guidance, recognition, enforcement), in which operators may run simultaneous approaches with different suppliers and return to activities as required, according to business impacts, suppliers, and products or destinations. Secondly, it proposes that SSCM decision making can be approached and analyzed through the lens of shared value creation (in terms of how value is created and for whom) and embracing paradoxical thinking to generate creative solutions to conflicting demands. These perspectives can be usefully applied in further study. The grounding of the framework concepts in literature from the field of tourism management, purchasing, and supply chain management, and further developed through interview analysis, provide a tool to carry out a global comparison of differing international operator approaches. This is of critical importance to developing insight into a sector that continues to be influential in shaping international tourism patterns and plays a particular role in instilling consumer confidence to travel in the period post-COVID-19 global lockdown.

SSCM approaches will differ according to company size/structure/ownership, resources, strategic outlook, sustainability expertise, and characteristics of company products, suppliers, and destinations. The framework presented here purposely does not include organizational features that influence the implementation of SSCM, as this would detract from the detailed focus on SSCM activities, but it is an important area for further research. Such influencing factors on SSCM activities can be usefully applied to further advance understanding, especially if linked to performance outcomes linked to value created for differing stakeholders. To develop the framework here, it would also be helpful to apply responsible management typologies (Dyllick and Muff, Silvestre) to the analysis of SSCM, to build more insights into the organizational characteristics required for transformational action.

While a strength of this research is its empirical basis in operators of differing scales, a limitation is that it does not capture the views of destination suppliers. A study that contrasted operator with supply chain perspectives would present a more robust view and uncover more perspectives and would also ideally extend to the entire supply chain including destination management organizations and end consumers. It is only through understanding each element of the supply chain ecosystem that the potential value of each player can be understood, alongside full insights into the paradoxical nature of decision making.

It is unlikely that full implementation of SSCM will be appropriate for every supplier relationship since projects require the development of trust between partners if a significant investment is to be made. Three conditions are necessary for meaningful SSCM – long-term partnership, fair pricing, and a consistent volume of operations. Characteristics of the tour operating industry such as high levels of price competition, fluctuating market demand, and a possible short-term outlook to business relationships pose a considerable threat to industry-wide engagement in SSCM. In response to the current troubled times of tourism, some destinations may revert to 'business as usual' commercial imperatives while others may potentially re-build economies through 'new' approaches that embrace finding business solutions that mutually serve people, the planet and profit goals. While the pandemic has threatened long term stability and planning, it has also deepened awareness of the interconnections in systems that are at the heart of SSCM and therefore, paradoxically, might be a positive force for good in building longer-term resilience to troubled times.

KEYWORDS

- **global tourism industry**
- **paradox theory**
- **shared value analysis**
- **supply chain management**
- **sustainable development goals**
- **tour operations**

REFERENCES

Al-Aomar, R., & Hussain, M., (2017). An assessment of green practices in a hotel supply chain: A study of UAE hotels. *Journal of Hospitality and Tourism Management, 32*, 71–81.

Association of Independent Tour Operators, (2020). *Travel Insights Report 2020.* https://www.aito.com/media-area/press-office/travel-insights-2020 (accessed on 20 January 2022).

Baddeley, J., & Font, X., (2011). Barriers to tour operator sustainable supply chain management. *Tourism Recreation Research, 36*(3), 205–214. https://doi.org/10.1080/02508281.2011.11081667.

Bastakis, C., Buhalis, D., & Butler, R., (2004). The perception of small and medium-sized tourism accommodation providers on the impacts of the tour operators' power in the Eastern Mediterranean. *Tourism Management, 25*(2), 151–170. https://doi.org/10.1016/s0261-5177(03)00098-0.

Black, R., Ham, S., & Weiler, B., (2001). Ecotour guide training in less developed countries: Some preliminary research findings. *Journal of Sustainable Tourism, 9*(2), 147–156.

Bocken, N. M. P., Rana, P., & Short, S. W., (2015). Value mapping for sustainable business thinking. *Journal of Industrial and Production Engineering, 32*(1), 67–81. https://doi.org/10.1080/21681015.2014.1000399.

Bocken, N. M. P., Short, S. W., Rana, P., & Evans, S., (2014). A literature and practice review to develop sustainable business model archetypes. *Journal of Cleaner Production, 65*, 42–56. https://doi.org/10.1016/j.jclepro.2013.11.039.

Bonilla-Priego, M. J., Nájera-Sánchez, J. J., & Font, X., (2021). Beyond ethics: The transformational power of overlapping motivations in implementing strategic sustainability actions. *Business Strategy and the Environment*. https://doi.org/10.1002/bse.2912.

Budeanu, A., (2009). Environmental supply chain management in tourism: The case of large tour operators. *Journal of Cleaner Production, 17*(16), 1385–1392. https://doi.org/https://doi.org/10.1016/j.jclepro.2009.06.010.

Buhalis, D., (2000). Relationships in the distribution channel of tourism: Conflicts between hoteliers and tour operators in the Mediterranean region. *International Journal of Hospitality and Tourism Administration, 1*(1), 113–139.

Carey, S., Gountas, Y., & Gilbert, D., (1997). Tour operators and destination sustainability. *Tourism Management, 18*(7), 425–431. https://doi.org/10.1016/s0261-5177(97)00044-7.

Crotty, M., (2006). *The Foundations of Social Research*. Sage Publications.

Curtin, S., & Busby, G., (1999). Sustainable destination development: The tour operator perspective. *International Journal of Tourism Research*, 135–147.

Curtin, S., & Wilkes, K., (2005). British wildlife tourism operators: Current issues and typologies. *Current Issues in Tourism, 8*(6), 455–478.

Font, X., English, R., Gkritzali, A., & Tian, W., (2021). Value co-creation in sustainable tourism: A service-dominant logic approach. *Tourism Management, 82*, 104200. https://doi.org/10.1016/j.tourman.2020.104200.

Font, X., Sanabria, R., & Skinner, E., (2003). Sustainable tourism and ecotourism certification: Raising standards and benefits. *Journal of Ecotourism, 2*(3), 213–218.

Font, X., Tapper, R., Schwartz, K., & Kornilaki, M., (2008). Sustainable supply chain management in tourism. *Business Strategy and the Environment, 17*, 260–271.

Forsyth, T., (1995). Business attitudes to sustainable tourism: Self-regulation in the UK outgoing tourism industry. *Journal of Sustainable Tourism, 3*(6), 210–230. https://doi.org/https://doi.org/10.1080/09669589509510727.

Goffi, G., Masiero, L., & Pencarelli, T., (2018). Rethinking sustainability in the tour-operating industry: Worldwide survey of current attitudes and behaviors. *Journal of Cleaner Production, 183*, 172–182. https://doi.org/10.1016/j.jclepro.2018.02.029.

Gössling, S., Scott, D., & Hall, C. M., (2021). Pandemics, tourism and global change: A rapid assessment of COVID-19. *Journal of Sustainable Tourism, 29*(1), 1–20. https://doi.org/10.1080/09669582.2020.1758708.

Hahn, T., Figge, F., Pinkse, J., & Preuss, L., (2017). A paradox perspective on corporate sustainability: Descriptive, instrumental, and normative aspects. *Journal of Business Ethics, 148*(2), 235–248. https://doi.org/10.1007/s10551-017-3587-2.

Hahn, T., Pinkse, J., Preuss, L., & Figge, F., (2014). Tensions in corporate sustainability: Towards an integrative framework. *Journal of Business Ethics, 127*(2), 297–316. https://doi.org/10.1007/s10551-014-2047-5.

Hall, C. M., Scott, D., & Gössling, S., (2020). Pandemics, transformations and tourism: Be careful what you wish for. *Tourism Geographies, 22*(3), 577–598. https://doi.org/10.1080/14616688.2020.1759131.

Hamid, M. A., Isa, S. M., & Kiumarsi, S., (2021). Sustainable tourism practices and business performance from the tour operators' perspectives. *Anatolia, 32*(1), 23–32. https://doi.org/https://doi.org/10.1080/13032917.2020.1830135.

Hartley, J., & Choi, T., (1996). Supplier development: Customers as a catalyst of process change. *Business Horizons,* 37–44.

Hines, P., (1994). Internationalization and localization of the Kyoroku air: The spread of best practice supplier development. *The International Journal of Logistics Management, 5*(1), 67–72.

Humphreys, P. K., Li, W. L., & Chan, L. Y., (2004). The impact of supplier development on buyer-supplier performance. *Omega, 32*, 131–143.

Kalisch, A., (2002). *Corporate Futures: Consultation on Good Practice*. Tourism Concern.

Khairat, G., & Maher, A., (2012). Integrating sustainability into tour operator business: An innovative approach in sustainable tourism. *Tourismos: An Interdisciplinary Journal of Tourism, 7*(1), 213–233.

Klemm, M., & Parkinson, L., (2001). UK tour operator strategies: Causes and consequences. *International Journal of Tourism Research, 3*, 367–375. https://doi.org/DOI: 10.1002/jtr.351.

Krause, D., & Scannell, T., (2002). Supplier development practices: Product- and service-based industry comparisons. *The Journal of Supply Chain Management, 38*(2), 13–21.

Lozano, J., Arbulú, I., & Rey-Maquieira, J., (2016). The greening role of tour operators. *Environmental Management, 57*, 49–61.

Mayer, C., & Roche, B., (2021). *Putting Purpose into Practice: The Economics of Mutuality.* Oxford University Press.

Medina-Muñoz, R. D., Medina-Muñoz, D. R., & García-Falcón, J. M., (2003). Understanding European tour operators' control on accommodation companies: An empirical evidence. *Tourism Management, 24*(2), 135–147. https://doi.org/10.1016/s0261-5177(02)00062–6.

Miller, G., (2001). Corporate responsibility in the UK tourism industry. *Tourism Management, 22*(6), 589–598. https://doi.org/10.1016/s0261-5177(01)00034-6.

Modi, S. B., & Mabert, V. A., (2007). Supplier development: Improving supplier performance through knowledge transfer. *Journal of Operations Management, 25*, 42–64.

Newman, R. G., & Rhee, K. A., (1990). A case study of NUMMI and its suppliers. *Journal of Purchasing and Materials Management, Fall*, 15–20.

Picazo, P., (2018). Tour operators' marketing strategies and their impact on prices of sun and beach package holidays. *Journal of Hospitality and Tourism Management, 35*, 17–28.

Porter, M. E., & Kramer, M. R., (2006). Strategy and society: The link between competitive advantage and corporate social responsibility. *Harvard Business Review*, 76–93.

Prahinski, C., & Benton, W. C., (2004). Supplier evaluations: Communication strategies to improve supplier performance. *Journal of Operations Management, 22*, 39–62.

Quayle, M., (2000). Supplier development for UK small and medium sized enterprises. *Journal of Applied Management Studies, 9*(1), 117–133.

Richards, P., & Font, X., (2019). Sustainability in the tour operator – ground agent supply chain. *Journal of Sustainable Tourism, 27*(3), 277–291. https://doi.org/10.1080/09669582.2018.1560451.

Schwartz, K., Tapper, R., & Font, X., (2008). A sustainable supply chain management framework for tour operators. *Journal of Sustainable Tourism, 16*(3), 298–314. https://doi.org/10.1080/09669580802154108.

Sigala, M., (2008). A supply chain management approach for investigating the role of tour operators on sustainable tourism: The case of TUI. *Journal of Cleaner Production, 16*(15), 1589–1599. https://doi.org/10.1016/j.jclepro.2008.04.021.

Sigala, M., (2020). Tourism and COVID-19: Impacts and implications for advancing and resetting industry and research. *Journal of Business Research, 117*, 312–321. https://doi.org/10.1016/j.jbusres.2020.06.015.

Smith, W. K., & Lewis, M. W., (2011). Toward a theory of paradox: A dynamic equilibrium model of organizing. *Academy of Management Review, 36*, 381–403.

Sriram, V., Krapfel, R., & Spekman, R., (1992). Antecedents to buyer-seller collaboration: An analysis from the buyer's perspective. *Journal of Business Research, 25*, 303–320.

Tan, K., Handfield, R., & Krause, D., (1998). Enhancing the firm's performance through quality and supply base management: An empirical study. *International Journal of Production Research, 36*(10), 2813–2837.

Tapper, R., (2001). Tourism and socio-economic development: UK tour operators' business approaches in the context of the new international agenda. *International Journal of Tourism Research, 3*(5), 351–366. https://doi.org/10.1002/jtr.348.

Tay, K. X., Chan, J. K. L., Vogt, C. A., & Mohamed, B., (2016). Comprehending the responsible tourism practices through principles of sustainability: A case of Kinabalu park. *Tourism Management Perspectives, 18*, 34–41. https://doi.org/10.1016/j.tmp.2015.12.018.

Tveteraas, S., Asche, F., & Lien, K., (2014). European tour operators' market power when renting hotel rooms in Northern Norway. *Tourism Economics, 20*(3), 579–594. https://doi.org/10.5367/te.2013.0291.

Walter, A., (2003). Relationship-specific factors influencing supplier involvement in customer new product development. *Journal of Business Research, 56*, 721–733.

Weiler, B., & Ham, S. H., (2002). Tour guide training: A model for sustainable capacity building in developing countries. *Journal of Sustainable Tourism, 10*(1), 52–69.

World Travel and Tourism Council, (2020). *To Recovery and Beyond: The Future of Travel and Tourism in the Wake of COVID-19.* https://wttc.org/Portals/0/Documents/Reports/2020/To%20Recovery%20and%20Beyond-The%20Future%20of%20Travel%20Tourism%20in%20the%20Wake%20of%20COVID-19.pdf?ver=2021-02-25-183120-543 (accessed on 20 January 2022).

World Travel and Tourism Council, (2021). *Lessons Learnt During COVID-19.* https://wttc.org/Portals/0/Documents/Reports/2021/Lessons-Learnt-%20COVID-19.pdf?ver=2021-08-19-095731-037 (accessed on 20 January 2022).

Xiao, C., Wilhelm, M., Van, D. V. T., & Pieter, V. D. D., (2019). Inside the buying firm: Exploring responses to paradoxical tensions in sustainable supply chain management. *Journal of Supply Chain Management, 55*(1), 3–20. https://doi.org/https://doi.org/10.1111/jscm.12170.

Yang, Y., Zhang, C. X., & Rickly, J. M., (2021). A review of early COVID-19 researches in tourism: Launching the annals of tourism research's curated collection on coronavirus and tourism. *Annals of Tourism Research, 91*, 103313. https://doi.org/10.1016/j.annals.2021.103313.

Zapata, C. M. J., Hall, C. M., & Backlund, S., (2018). Can MNCs promote more inclusive tourism? Apollo tour operator's sustainability work. *Tourism Geographies, 20*(4), 630–652. https://doi.org/10.1080/14616688.2018.1457074.

Index

D

M

N

O